THE WORLD OF THE HUMMINGBIRD

THE WORLD OF THE HUMMINGBIRD

ROBERT BURTON

FIREFLY BOOKS

A FIREFLY BOOK

Published by Firefly Books Ltd. 2001

Copyright © 2001 Robert Burton

All rights reserved. No part of this publication may be reproduced, stored in a retrieval system or transmitted in any form or by any means, electronic, mechanical, photocopying, recording or otherwise, without the prior written permission of the Publisher.

First Printing 2001

U.S. Cataloging-in-Publication Data
(Library of Congress Standards)

Burton, Robert
   The world of the hummingbird / Robert Burton. – 1st ed.
[160] p. : col. ill. ; cm.
Includes bibliographical references and index.
Summary: Introduction to the hummingbird, its flight, relationship with plants and humans, nesting behavior, migration, myth and legend, trade in hummingbird feathers, and current threats to its habitat.
ISBN 1-55209-607-6
1. Hummingbirds. I. Title.
598.764 21 2001

National Library of Canada Cataloguing
in Publication Data

Burton, Robert, 1941 –
   The world of the hummingbird

Includes index.
ISBN 1-55209-607-6

1. Hummingbirds. I. Title.

QL696.A558B87 2001   598.7'64   C2001-930614-8

Published in the United States in 2001 by
Firefly Books (U.S.) Inc.
P.O. Box 1338, Ellicott Station
Buffalo, New York 14205

Published in Canada in 2001 by
Firefly Books Ltd.
3680 Victoria Park Avenue
Willowdale, Ontario M2H 3K1

Produced by
Bookmakers Press Inc.
12 Pine Street
Kingston, Ontario K7K 1W1
(613) 549-4347
tcread@sympatico.ca

Design by
Robbie Cooke-Voteary and Janice McLean

Printed and bound in Canada by
Friesens
Altona, Manitoba

Printed on acid-free paper

The Publisher acknowledges the financial support of the Government of Canada through the Book Publishing Industry Development Program for its publishing activities.

Front cover: Violet sabrewing (*Campylopterus hemileucurus*)
© Michael and Patricia Fogden
Back cover: Crowned woodnymph (*Thalurania furcata*)
© Michael and Patricia Fogden

CONTENTS

> "Of all the numerous groups into which the birds are divided, there is none other so numerous in species, so varied in form, so brilliant in plumage, and so different from all others in their mode of life…without doubt, the most remarkable group of birds in the entire world."
>
> Robert Ridgway
> *The Hummingbirds (1890)*

FLYING JEWELS

Chapter One

I saw my first wild hummingbird at a cabin in Portal, Arizona, a hamlet nestled in the shadow of the Chiricahua Mountains. Our bags were still unpacked when the sound of whirring wings announced a hummingbird's arrival at the sugar-water feeder just outside the window. It was a black-chinned hummingbird (*Archilochus alexandri*), the most common of the local species and one I would come to know well.

We had come to Arizona especially to see hummingbirds, and the guidebook assured us that Portal was one of the best places to find them. We learned that nearby Cave Creek Canyon boasts sightings of 330 bird species, 14 of which are hummingbirds. Those 14 represent fully two-thirds of the number of hummingbird species ever recorded in the United States.

The Chiricahua Mountains rise

A crowned woodnymph (Thalurania furcata) *with a drop of nectar on the tip of its bill.*

A great sapphirewing (Pterophanes cyanopterus) *feeds on* Lobelia *in the forested slopes of the Andes in Colombia.*

from the desert floor in the south-eastern corner of Arizona, where that state abuts New Mexico to the east and Mexico to the south. It was in this neighborhood that Cochise and Geronimo led the Apaches' forlorn attempt to stem the tide of white settlement. With an elevation of more than 10,000 feet, the mountains offer habitats ranging from arid desert to cool pine forest, creating one of the most diverse landlocked plant and animal communities in the United States. Not surprisingly, these mountains are a prime birding site that attracts visitors from all over the country and beyond.

However, rather than join the daily procession that winds up the forest trails in search of the elusive but brilliant elegant trogon and other specialties, I was content to settle back and watch the hummingbirds. This was made easy by the generosity of the citizens of Portal, who not only set out an array of feeders in their gardens but put seats around them so that total strangers can savor the spectacle of the hummingbirds, one of nature's wonders.

Quite apart from the dazzling kaleidoscope of their colors and the sheer panache of their movement, hummingbirds are unrivaled among birds in several ways, which makes them particularly interesting to zoologists. They have a unique flight mechanism that enables them to hover at ease and even fly backward. They are the smallest warm-blooded animals and the only birds to go regularly into a form of hibernation at

"A glittering fragment of the rainbow... a lovely little creature moving on humming winglets through the air, suspended as if by magic in it, flitting from one flower to another, with motions as graceful as they are light and airy."

John James Audubon (1840)

A female broad-billed hummingbird (Cynanthus latirostis) *tends to her nestling.*

night by allowing their metabolism to slow and their body temperature to drop dramatically. They also have such physiological adaptations as the relatively largest flight muscles and heart, the highest rate of heartbeat and the densest plumage of any bird. Although diminutive, they are birds of superlatives. Eye-catchingly beautiful and exquisitely miniature, they are like no other family of bird in their intensity of life. Everything they do occurs at heightened speed and on blurred wings.

I have long been interested in bird flight, but my interest is not entirely in the technicalities of the wing-beat mechanisms of which hummingbirds are such supreme exponents. I am curious, instead, about the way birds use the power of flight in their daily lives; in other words, my fascination lies in the ecology of flight.

Flight provides birds with access to three dimensions and the ability to travel at speed. But flight has a drawback—its high energy cost. There are examples throughout the bird world of the measures birds adopt to reduce the heavy penalty they pay for the ability to fly. Vultures soar on thermals, and albatrosses glide on breezes blowing over the oceans. Even commonplace garden birds use energy-saving

A blue-throated hummingbird
(Lampornis clemenciae) *flashes*
its brilliant throat feathers while
feeding at Thistle Cave Creek, Arizona.

The rufous hummingbird (Selasphorus rufus) *migrates to nest as far north as Alaska.*

strategies to make flying less strenuous. The hovering flight of hummingbirds, however, is the most energy-demanding of any form of locomotion in the animal kingdom. Its advantage is that it gives hummingbirds access to nectar, an energy-rich food. In this sense, the flight style and energy source are dependent on each other; the relationship between flight and food is the key to understanding the lives of hummingbirds.

During the four days we spent at Portal, I spotted seven hummingbird species. Some were local specialties such as the magnificent (*Eugenes fulgens*), blue-throated (*Lampornis clemenciae*) and broad-billed (*Cynanthus latirostris*) hummingbirds; others were en route to more northerly nesting grounds. Identification was not always easy. New species had to be picked out from the swarm of hummers that buzzed around the feeders. Some paid only brief visits, either through natural shyness or because they were driven off by more aggressive birds. In fact, if a local birder had not pointed it out to me, I would not have noticed the one brief appearance of a calliope (*Stellula calliope*) hummingbird.

Another barrier to identification is the visual challenge created by the patches of shimmering iridescent plumage that make the diminutive hummingbirds such exquisite "flying jewels." It is one thing to learn the position and color of the iridescent feathers shown in the field guides, but unless the light strikes the plumage at the right angle, they do not always show up on the living bird. Often there is only a dark patch that may not even contrast with the surrounding feathers. The throat of the black-chinned hummingbird may look as black as the chin until a sudden movement catches the light and the throat is illuminated by a flash of violet iridescence. Then, just as suddenly, it disappears as if it has been switched off.

In Portal, I discovered it is easier to identify hummingbirds in the early morning, when the low sun strikes the birds at an angle that makes them iridesce at eye level. But there was never a problem picking out the brick-red plumage of the rufous hummingbird (*Selasphorus*

"Of all animated beings…the most elegant in form and brilliant in color… [nature] has loaded it with all the gifts of which she has only given other birds a share."

Comte de Buffon (1749)

rufus), which I could be certain of finding at a feeder beside a dry creek. The rufous became one of my favorites as I watched it defend the feeder against all comers, including the black-chinned hummingbird that normally ruled the roost and even the much larger magnificent and blue-throated hummingbirds. The rufous hummingbird was only passing through Portal, but I had watched it defend this site for two days. If any other hummingbird approached, the rufous would race toward it, tail fanned and squealing like a minuscule pig, to send the other bird veering away.

Our visit—timed to catch the hummingbirds on their migration northward—took place in April. I was so engrossed in watching hummers at the feeders that it took me some time to realize that there were no flowers in either the gardens or the surrounding countryside. The hummers relied wholly on sugar-water feeders, and it is these that make Portal such an excellent place to see migrant hummingbirds. Indeed, there is no need to walk in search of hummers. You choose your vantage point by a feeder and wait for them to appear. "Appear" is the right word. All too often, I did not even see them flying toward me—the humming sound would be the first notification of their arrival. Sometimes we could watch half a dozen or more at one time, and it was difficult to count them as they zoomed to and fro. Some species would spend time at one feeder; others fed briefly and departed. One apparent reason for a short

visit was aggression from other birds.

As it happened, I was observing an abnormal assembly of hummingbirds in Portal. The complete absence of flowers was the result of a particularly hard winter. In other years, many of the hummingbirds would be defending their own patches of flowers, but in the absence of flowers, they were competing for a place at the limited number of feeders. Without access to flowers, they were leading unnatural lives. The relationship between hummingbirds and the flowers that nourish them is the most important theme in the family's natural history. The evolution and ecology of hummingbirds are intimately linked to their reliance on nectar as a source of energy.

To learn more about hummingbirds, I searched the scientific literature for information that would help me understand and appreciate them. I soon discovered that they are popular subjects of study. I have no doubt that scientists enjoy watching hummingbirds for pleasure as much as anyone, but they have also found them very useful as research animals. Hummingbirds' ecology, use of habitat, migratory habits and breeding strategy have all been shaped by their feeding habits. Their capacity for sustained hovering flight, which sets them apart from other birds, including other nectar feeders, is made possible by radical adaptations in anatomy and physiology. Hummingbirds are, then, more than merely pretty birds to watch. Research into details of their habits can be used to unravel some of the mysteries of animal behavior and ecology.

A male coppery-headed emerald (Elvira cupreiceps) *settles briefly on an epiphytic orchid* (Maxillaria fulgens), *a flower visited mainly by bees.*

One subject that has been widely studied is the function of territory. A territory is a defended space, and we know a great deal about breeding territories in which birds, both male and female, defend their nests and the area nearby. But hummingbirds also defend territories that are purely for feeding. In so doing, they are guarding a food resource for their exclusive use. By counting the number of flowers a hummingbird visits, researchers have shown that it will adjust the size of its territory so that it always defends roughly the same number of flowers. If some of the flowers visited are enclosed in bags, the hummingbird compensates by expanding its territory to include additional flowers. When the bags are removed from the flowers, the territory shrinks to its former size.

During my stay in Portal, I saw how one hummingbird will defend a feeder against all comers. While the feeder provided the rufous hummer with a secure and exclusive source of nourishment, it was at the expense of the time and effort needed to drive other birds away. As there was enough sugar-water for all the hummingbirds, it seemed a pointless occupation. Nevertheless, defending a source of food is common behavior in hummingbirds as well as in many other animals, and biologists are fascinated by the rules of engagement: Who wins the contest, and under what circumstances are contests worth the effort? In the mid-1980s, Paul Ewald, then of the University of Michigan, studied this by setting up a line of feeders that delivered sugar

solution at two different fixed rates, 8 and 10 milliliters per hour. When two neighboring feeders had been adopted as territories by black-chinned hummingbirds, Ewald gradually moved them closer and closer together until the hummingbirds came into conflict. Then he watched to find out whether the hummingbird with the richer source of food was the victor in aggressive encounters. It was, because it had more to lose and so was more vigorous in its defense.

Simple, elegant experiments such as these, which do not harm or disturb the hummingbirds, have been used to shed light on many aspects of their behavior and ecology. For the purposes of this book, they allow a portrait of hummingbirds to go beyond a description of their natural history to an understanding of how these very special little birds survive and make a living in ways that at first seem astounding. From the point of view of the researchers, the value of the hummingbirds is that they are "tools" for investigating biological theories. They are confiding and relatively sedentary birds that are easy to observe, and their food supply can be monitored accurately, either by extracting nectar from plants and analyzing it by volume and sugar content or by providing feeders with metered amounts of sugar-water.

Researchers choose species of hummingbirds that are particularly easy to work with, often because they are abundant and live conveniently near a research establishment. So the names of some species—rufous, black-chinned,

A mass of Heliconia *plants exploits a light gap in the Costa Rican rainforest, creating feeding opportunities for a variety of hummingbird species.*

Anna's (*Calypte anna*) and sparkling violetear (*Colibri coruscans*)—crop up frequently in scientific studies, while even basic accounts of most of the 300 or so species, especially those living in the remoter parts of Central and South America, are still unwritten. In North America, hummingbirds are now among the best-studied group of birds, not only because they are good study animals for scientists but because of the general public's fascination with them. Despite this, many details of the lives of even common species—courtship, nesting habits and migration routes, for instance—are still not well understood, but the situation is changing as groups of hummingbird enthusiasts are beginning to amass and exchange information.

RANGE

Hummingbirds range from Alaska and Labrador in the north to Tierra del Fuego in the south and from Juan Fernandez Island in the west to Barbados in the east. They are unique to the Americas, never having managed to spread into the Old World by landward expansion across the Bering Strait or by chance transoceanic wanderings. Their habitats range from lowland rainforest and coastal mangrove swamps to desert, subarctic meadows and the snow line of the Andes.

Not surprisingly for birds that depend on flowers, hummingbirds are abundant in the Tropics, where plant growth is at its most exuberant, and are most varied in the Andes. The greatest concentration of

In the summer months, hummingbirds can be found as far north as Alaska, where the nectar of wild fireweed serves as one of their food sources.

species is in Colombia and Ecuador (about 135 and 163, respectively). Fewer species exist in the Amazon basin to the east: Brazil claims only 84 species. However, in relation to their small size, Central American countries—Costa Rica, Belize, Panama and El Salvador—have just as great a diversity.

The number of hummingbird species also decreases with altitude above sea level and distance from the equator. Only a few specialized forms live beside the glaciers in the Andes and in the northern and southern extremities of the Americas. One species, the rufous hummingbird, reaches Alaska, and another, the green-backed firecrown (*Sephanoides sephanoides*), lives in Tierra del Fuego, where Charles Dar-win watched it flitting about in snowstorms. British ornithologist G.T. Corley Smith found Chimborazo hillstars (*Oreotrochilus chimborazo*) nesting in the moorland páramo zone 15,000 feet up Mount Cotopaxi in Ecuador, where both the bird-watcher and the hummingbirds had to contend with daily hail showers and nightly frosts.

For the majority of eastern North Americans, the hummingbird most likely to be seen is the ruby-throated (*Archilochus colubris*); for westerners, it is the rufous. Only in the south and southwest is there a greater range of species. Fifteen hummingbird species nest in the conterminous United States (not including the record of a Xantus' hummingbird (*Hylocharis xantusii*) that once laid

Left, a long-tailed hermit (Phaethornis superciliosus) *visiting a* Heliconia *flower. Right, a male Anna's humming-bird* (Calypte anna) *in flight.*

eggs in California but did not hatch them); 5 of the 15 nest in Canada. Eight are confined to the Mexican border states of Arizona, New Mexico and Texas. A further eight species wander north of the Mexican border, some regularly, others infrequently, but do not nest in the United States. These vagrants are usually seen between California and Texas, but West Indian species sometimes appear in Florida, and there is one amazing record of a Xantus' hummingbird spotted in British Columbia. This species had not previously been seen north of California, and it is a rarity even there.

NAMES

The brilliance of the hummingbirds' plumage is paralleled by their exotic names: Jamaican mango, festive coquette, shining sunbeam, empress brilliant, hyacinth visorbearer, marvelous spatuletail, and many more. Not surprisingly, precious stones appear in many hummingbird names: topaz, sapphire, ruby, emerald, tourmaline and beryl. Because they are so colorful, even basic descriptive names are evocative—heliotrope-throated, indigo-capped, fork-tailed woodnymph and blue-tufted starthroat. However, beautiful hummingbird, charming hummingbird and magnificent hummingbird are unimaginative names that could be applied to many species.

Some hummingbird names are known to have been concocted by cage-bird dealers who either did not know the correct names of the birds in their stock or wanted to make their wares sound more attractive. This is one reason why some species have a confusing variety of common names. For instance, *Phaethornis superciliosus* has been called, at one time or another, long-tailed hermit, buff-browed hermit, Guiana hermit, allied hermit, Cayenne hermit and superciliosus. If nothing else, this shows the need for using the internationally accepted Latin-based scientific names to eliminate the confusion of common names.

Many hummingbird species were first described and given names when they were imported as dried skins (see Chapter 7) by so-called "cabinet naturalists," people who studied specimens in collections. Striking color patterns and unusually shaped bills or plumes were the only characteristics available for describing them. Occasionally, hummingbirds could be named for their place of origin, as were the Honduras emerald (*Amazilia luciae*) and the Ecuadorian piedtail (*Phlogophilus hemileucurus*). Others were named to honor or flatter friends or patrons. Anna's hummingbird was named for

A Costa's hummingbird (Calypte costae) *visits the fiery blossoms of a Californian autumn sage* (Salvia greggii).

the wife of Victor Masséna, Duke of Rivoli, who lived in the early 1800s. He was a keen collector of birds, and the name was given to a specimen in his collection (all of which was later bought by the Academy of Natural Sciences in Philadelphia). The Duke also gave his name to Rivoli's hummingbird (now known as the magnificent hummingbird). Costa's hummingbird (*Calypte costae*) was named for another French nobleman, Louis Marie Costa, Marquis de Beau-Regard. As the trade in bird skins was mostly transatlantic, the European aristocracy is well represented among hummingbird names, although Allen's hummingbird (*Selasphorus sasin*) is named for Charles Andrew Allen, a naturalist who spent part of his life in California, where this species lives. Xantus' hummingbird is named for John Xantus, the 19th-century bird collector who discovered it in Baja California.

CLASSIFICATION

According to the majority of ornithologists, hummingbirds share a common ancestor with the swifts. Ornithological classification recognizes this by placing the hummingbirds (family Trochilidae) with the swifts (Apodidae) and the three species of crested swifts (Hemiprocnidae) of southeast Asia in the order Apodiformes. At first glance, hummingbirds and swifts are very different. The swifts fly on long, slender wings designed for energy-saving flight that enables them to remain airborne for

Although the swifts' appearance and habits clearly distinguish them from the hummingbirds, they are close relatives. Shown here is the white-throated swift (Aeronautes saxatilis).

Revealing the structure of its wing, a characteristically dull-plumaged little hermit (Phaethornis longuemareus) hovers before an epiphytic heath.

weeks or months at a time while they coast around the open airspace in search of insects. By contrast, the hummingbirds burn up energy hovering in front of flowers but spend long periods perching.

However, important anatomical similarities between the two groups exist. In both birds, the wing bones have similar proportions: short upper-arm (humerus) and forearm (radius and ulna) bones but very long hand bones. The legs are short and weak, and most hummingbirds can do no more than shuffle along a perch (hence Apodiformes, which literally means "the legless ones"). On the other hand, the swifts fly the way other birds do, not like hummingbirds, and there are significant differences in breeding biology. In swifts, the sexes share the parental duties of incubating the eggs and feeding the young, but with rare exceptions, parental duties are left wholly to the female in hummingbirds. Another school of thought suggests that these and other differences are enough to separate the two groups and that hummingbirds are actually more closely related to the perching birds (Passeriformes), while the swifts are nearer the goatsuckers (Caprimulgiformes).

The difficulty in understanding the relationships among any groups of birds is that very few fossils exist to show how the different kinds of birds evolved and to provide the "missing links" between them. In recent years, DNA analysis has helped answer these questions. It is now believed that the hummingbirds and the swifts separated from their common ancestor approximately 57 million years ago in the Paleocene epoch and started to evolve into their main groups 40 million years ago in the Miocene.

There are two main groups of hummingbirds: the hermits and all the others (the "typical" hummingbirds). The hermits, which are sometimes given their own subfamily, Phaethornithinae, are dull-plumaged birds living in the Amazon basin; the sexes look alike. The "typical" hummingbirds, the Trochilinae, are much more numerous and probably evolved by colonizing the many new habitats that arose when the Andes and other American mountain ranges were formed.

Something like 328 species of hummingbird belong to the family Trochilidae. It is impossible to be precise about the numbers, because species can become extinct unnoticed or are rediscovered after they are presumed to have died out. There are also almost certainly species as yet undiscovered in remote wilderness areas of Central and South America, which are still difficult for ornithologists to reach. It is quite easy to overlook a species with a very restricted range such as the marvelous spatuletail (*Loddigesia mirabilis*), which lives in only one valley in Peru. As recently as 1996, Gary Stiles, a leading hummingbird expert, discovered the Chiribiquete emerald (*Chlorostilbon olivaresi*), which lives in the Sierra de Chiribiquete, in southeast Colombia, where it is the most common hummingbird.

To add to the challenge, ornithologists who classify birds do not always agree as to whether two very similar hummingbirds are separate species or variants of a single species. The question is complicated by the ease with which closely related hummingbirds interbreed. Hybrids between a number of species have been recorded. They are sometimes classified as new species because the hybrids have a mixture of characteristics from both parents and look like completely different birds. Hybrids are known between Anna's hummingbird and the cal-

The red-tailed comet (Sappho sparganura) *is native to central South America.*

liope hummingbird despite their difference in size. The puffleg hummingbird (*Eriocnemis dyselius*) of the Andes, which was described in 1872, is probably no more than a melanistic (black-pigmented) version of another species, *E. cupreoventris*.

Their nectar-drinking and hovering lifestyle has imposed a basic body form on the hummingbirds, making them as instantly recognizable as owls or penguins. There is considerable variation in size, from the bee hummingbird (*Calypte helenae*) of Cuba, which measures 2¼ inches long (half of which is bill and tail) and weighs 0.07 ounce, to the giant hummingbird (*Patagona gigas*) of the Andes, which is 8½ inches long (about the size of a European starling), weighs 0.7

ounce and has a 1-foot wingspan.

The main differences between species are the plumage and the length and shape of the bill. Bill length varies enormously, from the ⅜-inch version of the purple-backed thornbill, whose scientific name is *Ramphomicron microrhynchum* (translating literally as "tiny bill, tiny bill"), to the 4-inch bill of the sword-billed hummingbird (*Ensifera ensifera*). The greatest variation is in the plumage, however, especially in the males, who use their gaudy throat patches, called gorgets, as signals to rival males and prospective mates. By contrast, the females are usually drab, although in the hermits, violetears and emeralds, the females are similar to the males. It is the brilliance of their colors that has led to

A male black-chinned hummingbird (Archilochus alexandri) *shows the violet edge of its black throat.*

such rapturous descriptions of hummingbirds as "living jewels." John James Audubon described them as "glittering fragments of the rainbow." It is often unacknowledged, however, that some species of hummingbirds, including the males, are rather drab; one is even called the somber hummingbird (*Aphantochroa cirrochloris*). The brilliance of the plumage is enhanced in a few species by elaborate tail streamers or crests, but the great glory of hummingbirds is the glittering iridescent patches of feathers, particularly on the crown and neck.

This illustration of the structure of a hummingbird feather shows how the barbules are packed with light-splitting platelets. (After Johnsgard, 1997.)

IRIDESCENCE

The words of Alexander Wilson, the father of American ornithology, sum up this feature of hummingbird plumage, which is at once a delight and a problem for birders, as I found during my stay in Portal. Hummingbird feathers have two basic colors—black and rufous—produced by pigments in the feathers. Overlying these are the iridescent colors that change from moment to moment as the bird moves around the flowers and catches or loses the sun. Iridescent colors are "structural" colors formed by the microscopic internal arrangement of the feathers. Many birds such as ducks, jays, grackles and peafowl have bright patches of plumage caused by structural colors rather than pigments, but none glitter and shimmer as those of hummingbirds do.

Iridescence is caused by "interference coloration," the phenomenon that produces the rainbow colors on a

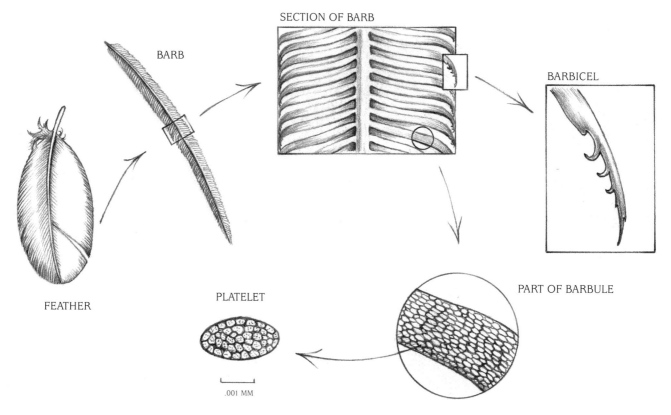

FEATHER

BARB

SECTION OF BARB

BARBICEL

PART OF BARBULE

PLATELET

.001 MM

The purple-backed thornbill's scientific name Ramphomicron microrhynchum *translates literally as "tiny bill, tiny bill."*

"What heavenly tints in mingling radiance fly!

Each rapid movement gives a different dye;

Like scales of burnished gold they dazzling show –

Now sink to shade, now like a furnace glow!"

Alexander Wilson (early 1800s)

soap bubble or a film of oil on a puddle in the road, in which the colors vary according to the angle at which light strikes the oil. In a feather, the iridescent colors are created by stacks of microscopic plates whose ordered structure produces a much more brilliant effect than a thin film of soap or oil does. In a hummingbird feather, iridescence is confined to the top third. The rest of the feather is devoted to providing insulation and streamlining and is, in any case, overlapped and covered by neighboring feathers. The vane of a feather is strong yet flexible, because the two rows of barbs on each side of the central quill that make up the vane divide again into two rows of thousands of tiny barbules linked by tiny hooks called barbicels. In the iridescent part of the feather, the barbules are not hooked together but are twisted so that their flat surfaces face the observer.

The physics of iridescence is not easy to understand. Briefly, iridescent colors are produced by a beam of light that is partly reflected and partly refracted, or bent, as it enters a thin layer of a transparent medium such as a film of soap or oil and then is reflected again at the back of the medium. Sunlight is split into its constituent colors, as in a rainbow, but the colors that are reflected depend on the thickness and the nature of the medium. In hummingbird feathers, the refractive medium is the layers of minute, air-filled plates in the barbules that break up light the way soap bubbles do. The thickness of the plates and the amount of air in them determine the degree of refraction and give hummingbirds colors that range through the spectrum from red to blue. Each color appears only when the barbules reflect light at a specific angle made by the observer, the bird and the sun. A slight change of angle, and the color changes or disappears completely and leaves the feather black.

Depending on the structure of a feather, its iridescence can be seen from a narrow range of angles or from almost any direction. Gorget feathers iridesce over a narrow angle, so the colors come and go as the bird moves. On other areas of plumage, especially the back, the iridescence is reflected over a wide range of angles, and the color appears to be permanent.

THE ULTIMATE FLIERS

Chapter Two

Hummingbirds' flight is unique among birds. It is sufficiently distinctive that a comparison of flight style can be made between hummingbirds and all other bird species. The main difference is that hummingbirds are the only birds which can hover for a prolonged period in still air. In the laboratory, a hummingbird has been persuaded to hover continuously for as long as 50 minutes, although in typical situations, they hover in short bouts of a few seconds to drain a flower.

Other birds can hover, usually with some difficulty, for no more than a few seconds. Small birds such as chickadees and thrushes hover briefly to gather food that is difficult to reach from a perch, such as an insect on a leaf or a berry hanging from a stalk. Larger birds appear to hover over open spaces for long periods while they search for food.

Above left, a sooty, or buff-tailed, barb-throat (Threnetes niger) *steals nectar, while a scintillant hummingbird* (Selasphorus scintilla) *shows how the wings turn over on the upstroke, facing page.*

Hawks, owls and jaegers use this form of hunting over land, and terns, ospreys and kingfishers seem to hover over water. The habit is a specialty of kestrels and black-shouldered kites. But none of these birds hovers in the strictest sense. They hang, apparently motionless, in one place, but they are, in fact, flying or gliding slowly into the wind so that their airspeed is equal and opposite to the wind speed, which makes their ground speed zero. This is known as wind-assisted hovering, or wind-hovering, and is impossible to achieve in a flat calm.

Flying by flapping wings is a strenuous, energy-demanding enterprise. To use flight as a means of locomotion, an animal has not only to propel itself forward but also to counteract gravity and prevent itself from falling out of the sky. This is in contrast to most water animals, which are buoyant and do not need to swim continually to keep themselves from sinking.

Many birds have found ways of reducing the effort of flight. Whenever possible, they "freewheel" by gliding. Birds such as albatrosses and vultures have taken this to an ex-

treme by remaining airborne without flapping for long periods. Hummingbirds, however, have evolved in the opposite direction and specialize in hovering, which is the most strenuous activity in the animal kingdom. Its reward is that they can station themselves with great precision in front of flowers while they drink the energy-rich nectar. If the flowers are swaying in the wind, hummers can fly forward, sideways and backward to maintain their position. They can even spin on their axes, turn over and fly on their backs when they need to escape in a hurry.

In normal flight, birds gain some lift from the flow of air over the wings that forward passage generates, and in wind-hovering, the wind blowing over the wings provides lift. In true hovering, all the lift is generated by the wings sweeping through the air and requires a rapid wing beat to keep the bird airborne. It is not easy to imagine such fast movements and harder still to see them. When a hummingbird is flying from place to place, it moves so quickly that it is difficult to keep track of it, let alone identify it, before it hurtles around a corner and out of sight. The opposite

A male coppery-headed emerald (Elvira cupreiceps) *feeds at a hotlips plant* (Cephaelis elata) *in the Costa Rican rainforest.*

is true when it is hovering. Closely approaching a hummingbird is easy, but the wings are a blur. Seen in profile, the bird looks as if it has two pairs of wings like a giant bee. This is because the wings become visible at the top and bottom of each wing beat as they slow down, stop and reverse direction. The shape of the wing beat can also be made out in the blur of movement.

The hovering flight of hummingbirds cannot be appreciated properly without understanding the basics of aerodynamics and how birds fly in straight and level flight. Until these principles were discovered, it was believed that birds somehow "rowed" through the air, pushing themselves forward and upward with sweeps of their wings. In fact, the wings are swept forward so that air flows over them and generates a lifting force. When air flows over the upper surface of a wing, it is forced to speed up and its pressure drops. The reverse takes place on the underside, where the air slows down and pressure rises. The result is that the wing is pulled up from above and pushed up from below. While the inner part of the wing, which bears the secondary flight feathers, has a simple up-and-down motion and generates lift continuously, the outer part, consisting of the wrist and fingers, which bear the primary flight feathers, is angled so that

the lift generated on the downstroke draws the bird forward.

By contrast, when hovering, a hummingbird's wings turn over on the upstroke so that they provide lift throughout the wing-beat cycle. This is not unlike the operation of a helicopter—hummingbirds' wings, like a helicopter's rotors, are swept through the air to provide continuous lift. The main difference is that nature has failed to invent the wheel, so hummingbird wings cannot revolve but have to sweep back and forth. A better analogy is the boatman who uses a single oar to scull from the back of his boat, swinging it from side to side and turning the blade so that water is forced backward and the boat forward with each stroke. The reciprocating wings and oar are less efficient than the spinning rotor because they have to slow down, reverse direction and accelerate at the end of each stroke.

The hummingbird wing is locked at the elbow and wrist to make a relatively inflexible paddle. When the hummingbird hovers, its body hangs almost upright so that the wings sweep back and forth in a horizontal plane, the wingtips describing a narrow figure-eight. The downstroke, or forward stroke, generates lift in the normal way, then the wings flip over at the shoulder, and the upstroke, or backstroke, strikes the air at the

Hovering, bottom to top: Wings sweep forward, rotate and return inverted, giving equal lift on up- and downstroke.

The spread tail of the coppery-headed emerald (Elvira cupreiceps) *gives the hummingbird extra lift.*

same angle in the opposite direction and continues to generate lift. As the lift is the same on both up- and downstrokes, the hummingbird hangs in the air.

Among birds, hummers are unique in their ability to generate as much lift on the downstroke as on the upstroke, but their flight is much more like that of insects, which are good at hovering and also have inflexible wings which describe a figure-eight motion. In the normal forward flight of other birds, the downstroke is the power stroke and the upstroke is little more than a "recovery" stroke to return the wing to position for the next downstroke. The supracoracoideus muscles responsible for raising the wings are correspondingly only a small fraction of the size of the pectoralis muscles that drive the downstroke. But in hummingbirds, where lift is generated throughout the wing-beat cycle, the supracoracoideus muscles are as much as half the size of the pectoralis, so the upstroke is correspondingly more powerful.

There are two occasions on which other birds use a wing-beat pattern similar to the hummingbird's figure-eight for hovering: the takeoff and the landing of a heavily built bird such as a dove. Like the hovering hummingbird, it has to remain airborne and under control when it is moving too slowly to generate lift by forward movement through the air. As it takes off, it crouches then jumps into the air with its body almost vertical (as is the hummingbird when it is hovering) so that the downstroke sweeps the wings forward and backward, rather than up and down, to create an airstream that will generate enough lift to keep the bird airborne. On the upstroke (now a backstroke), the outer half of the wing flips over, and the primary feathers separate, each one acting as a tiny wing to create lift. So much lift is generated that the bird rises almost vertically, but the effort is so great that the bird cannot sustain it for more than a few seconds.

Photographs of hovering hummingbirds often show the tail spread like a fan. This is a device used by many birds to get extra lift when they are flying slowly. Watch a large bird such as a crow or pigeon when it is about to land. The fanned tail acts as a third wing. For a hovering hummingbird, the extra lift is created by the airstream from the wings passing over the tail.

It is often said that hummingbirds have very fast wing beats to provide enough lift for hovering. The blurred wings and the hum reinforce the notion. The ruby-throated hummingbird is often cited as having a wing beat of 80 bps (beats per second) while hovering, which is much faster than in other birds. A chickadee, for instance, has a wing beat of about 25 bps.

But simple numbers can be misleading. A bird's wing-beat rate depends on its wing length. A heron has a wing beat of about 2 bps and the much smaller mockingbird one of 14 bps. This is the result of the basic mechanical law of oscillation. The wings act rather like the pendulum of an old-fashioned clock, the flight muscles taking the place of the clockwork mechanism. The length of the clock pendulum is adjusted so that it swings through a complete arc in exactly one second. It is then said to oscillate at a "natural frequency" of one second. If the weight on the pendulum is moved nearer the fulcrum from which it swings, the pendulum speeds up; if the weight is moved farther away, the pendulum slows down. This can be demonstrated easily by swinging a weight on the end of a string.

Birds' wings have a similar natural frequency and maintain a constant wing beat that depends on their length. As the theory of natural frequency and the comparison with a pendulum show, it is only to be expected that hummingbirds with tiny wings have faster wing beats than do birds with longer wings. In turn,

Compared with most other birds, hummingbirds have rigid wings with short arm bones but long hand bones.

HUMMINGBIRD

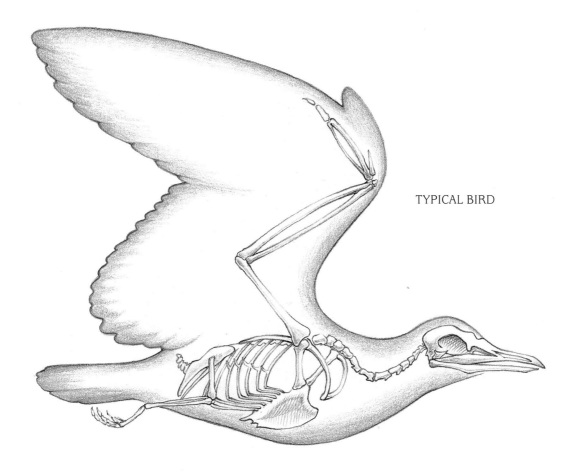

TYPICAL BIRD

hummingbird wing beats are slow by comparison with still smaller insects.

Crawford Greenewalt, who recorded hummingbird flight with high-speed photography in the 1950s, observed a wing-beat rate for a female ruby-throated hummingbird of a steady 53 bps, plus or minus 3. It was Greenewalt who realized that the constancy of birds' wing beats obeyed the law of oscillation and that hummingbirds have rapid wing beats as a result of their diminutive size. He also pointed out that hummingbirds would be expected to have slower wing beats than other birds in relation to their size.

This was confirmed by more pioneering studies by Walter Scheithauer, who used a stroboscope (an instrument that delivers very brief, rapid flashes of light) to investigate hummingbird wing beats. If the rate of flashing is synchronized with a bird's wing beats, the light will illuminate the wings only when they are in the same part of the cycle. They will appear motionless, and the bird will seem to hang in the air. Scheithauer's observations show that most hummingbird species have wing-beat rates of 20 to 25 bps, smaller species reaching 70 to 80 bps. The largest hummingbird, the giant hummingbird of the Andes, beats its wings at 8 to 10 bps, which is not much faster than the similar-sized European starling, and in forward flight, it alternates flapping with gliding as a starling does. Moreover, hummingbirds with rates of 20 to 25 bps are half the weight of chickadees with comparable rates. So weight for weight, hummingbirds actually beat their wings at about the same speed as, or more slowly than, other birds of comparable weight. The explanation for this paradox is given by the hummingbird's unique wing-beat mechanism. For the most part, other birds generate lift only on the upstroke and must flap their wings faster to get the same amount of lift as a hummingbird, which generates lift on both upstroke and downstroke.

Nowadays, there is no need to bring hummingbirds into the laboratory to measure their wing beats. The "hum" of free-flying wild birds can be recorded on tape and analyzed at leisure because the pitch of the hum indicates the wing-beat frequency. This technique has made it possible to determine the wing-beat speeds of many species. The distinctive whistle of the rufous hummingbird is produced by wing beats of 60 to 65 bps, while Anna's hummingbird, which is larger, beats its wings more slowly, at 45 bps. In their display flights, male rubythroats produce an incredible 200 bps.

With wings spread, a black-crested coquette (Paphosia helenae) *shows off the plumes on its head that can be raised in display.*

Nectar passes through the humming-bird's digestive system in a matter of minutes, as shown in this magenta-throated woodstar (Philodice bryantae).

ENERGY REQUIREMENTS

Size is important for hummingbirds' hovering lifestyle. Their light weight gives them the surplus power needed for hovering with ease. Every bird is limited in its capability for flight by the amount of power its breast muscles can deliver to its wings. There are two levels of power: The first is what the bird can produce on a sustainable basis to maintain its daily activities or use for long-distance flights when migrating; the second is the peak power it can deliver for a short time when taking off or escaping a predator. For many birds, the power they need to get off the ground at takeoff exceeds their sustainable power, and they can hover for only a moment. A California condor is so heavy that takeoff requires more power than the flight muscles can produce, and it can get airborne only by dropping from a perch or running to take off, just as an airplane gathers speed on the runway. At the other end of the scale, a hummingbird is so light, its sustainable power is greater than that needed for takeoff, and it has power to spare for continuous hovering. So much surplus power is available that a female hummingbird needing to change her position on her eggs does not stand and shift around like other birds but simply opens her wings, lifts off, swings round and settles again.

The diminutive size of hummingbirds, their skill at hovering and their diet of nectar are interlinked. A bird so small has to live at high speed and needs the energy-packed, easily assimilated diet that nectar can supply. The other side of the coin is that only a small animal can find sufficient nectar to satisfy its needs. Hovering gives hummingbirds access to nectar, but hovering is the most energy-expensive form of locomotion, and despite the fact that hummingbirds spend most of their time perching, they have to consume up to one and a half times their body weight of nectar every day. To process this vast amount of fluid, they have weak kidneys that rapidly excrete the excess water. The rich diet is also needed because they have a very high metabolic rate.

Metabolic rate (the rate of energy output relative to body weight) is the speed of the basic body processes—the chemical reactions that provide energy for heat, movement, growth, repair and reproduction. The

energy is obtained by the oxidation of food (carbohydrates, fats and proteins) and can be measured by tracking either the amount of food the animal eats or the amount of oxygen it requires to oxidize the food and release its energy.

This may seem a rather remote and technical subject, although it is important for anyone attempting to lose weight by diet or exercise. But it is important because a knowledge of the metabolic regime of hummingbirds helps answer questions about their daily activity, their need for frequent meals and the remarkable ability of some species to migrate hundreds of miles nonstop. It also highlights the dangers posed by even a brief interruption in their food supply. Moreover, the energy requirements of free-living hummingbirds can be measured much more easily than those of most animals because of their reliance on nectar. In study situations, they can be supplied with sugar-water of a known concentration in a feeder, or the nectar content of the flowers they visit can be analyzed. The ease with which their nutrition can be measured, whether they live in the wild or are in controlled laboratory conditions, is one reason why hummingbirds make such good study animals.

Hummingbirds are once again record holders—they have the highest metabolic rate of any warm-blooded vertebrate and, in the entire animal kingdom, take second place only to some large insects. They burn up so much energy for two reasons. The smaller the body, whether animal or mineral, the greater its surface area relative to its weight and the faster it will lose (or gain) heat. Thus hummingbirds need a high metabolic rate simply because of their small size, and as the smallest warm-blooded animals, they will have the fastest metabolism and must oxidize large quantities of carbohydrates to maintain their body temperature. The basic metabolic rate (when an animal is resting quietly and has no meal to digest) of a hummingbird is 12 times that of a much larger pigeon.

The whirring, hovering flight adds to the demands on hummingbird metabolism. With the unique wing-beat pattern that generates lift on both strokes, hummingbirds need a powerful "engine." Without airflow from forward movement to augment the lift provided by the beating wings, a hovering hummingbird has to work extra hard to stay airborne. Its wing muscles are not only much larger than those of other birds, but weight for weight, they also generate more power. The wing muscles make up a quarter to a third of the bird's entire body weight (compared with a seventh to a quarter in other birds) and are packed with a uniquely dense concentration of mitochondria, the microscopic bodies in every tissue cell that are responsible for energy release.

To keep the flight muscles and the rest of the body working at such a high pitch, hummingbird bodies need to circulate copious amounts of carbohydrates and oxygen. A hummingbird increases its oxygen intake sevenfold when it moves from quietly perching to hovering.

A stripe-tailed hummingbird (Eupherusa eximia) *hovers to collect nectar.*

In order to supply carbohydrates and oxygen to the muscles for this activity, hummingbirds have a very efficient circulatory system. Their heartbeat is rapid; the figures of 1,260 beats per minute in flight and 480 at rest for a blue-throated hummingbird are often quoted.

Neither figure is very different from the heartbeat rate of a chickadee, however. It is more important that the entire system is extremely effective. Hummingbirds' oxygen-transport system is much better developed than that of other birds. Their lungs have an outstanding capacity for absorbing oxygen from air, and their blood is richly supplied with red blood corpuscles that transport oxygen from the lungs to the rest of the body. The blood is pumped by a relatively large heart (up to 2.4 percent of the body weight, compared with 1.4 percent for a pigeon and less than 1 percent for an ostrich) and sends a copious flow of blood to the muscles, where well-developed networks of capillaries distribute it among the fibers.

To keep this high-speed mechanism running, hummingbirds must have a continuous and plentiful supply of food. At one time, it was believed that they consumed twice their own weight of sugar per day. This notion was based on a faulty

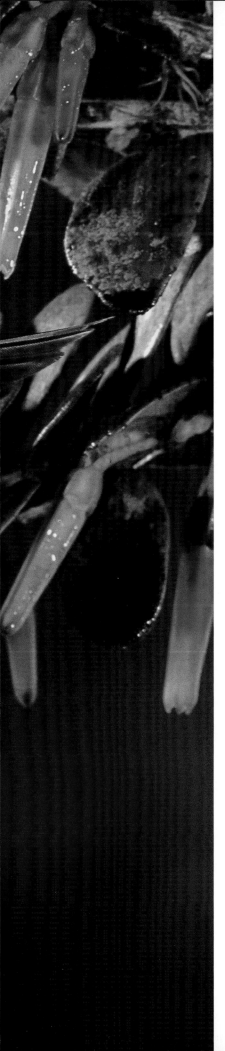

observation—the true figure is nearer half the body weight. Once again, this high figure is a direct result of hummingbirds' small size. Comparative figures are 30 percent of body weight for a chickadee and 3 to 4 percent for a chicken. If hummingbirds were any smaller, they would not be able to eat fast enough to avoid starvation—their metabolic rates would simply be too high for them to maintain the necessary supply of carbohydrate even though it is easily obtained as nectar.

The problem of processing so much food is eased by the nature of hummingbirds' diet. As long as plenty of nectar-rich flowers are available, a hummingbird does not spend an excessive amount of time feeding. Nectar, being a solution of sucrose, is quickly digested, and hummingbirds assuage their huge hunger with frequent small snacks throughout the day. There are usually no more than a few minutes' rest between each feed while the crop, a pouch in the throat for storing food, is quickly emptied. The shape of the stomach, with the entrance close to the exit, allows nectar to pass through quickly and enter the intestine, where it is rapidly digested. The whole process often takes less than 15 minutes.

One advantage of this system is that hummingbirds are not ham-

The male and female green violetear (Colibri thalassinus) *are alike in appearance, an unusual trait in hummingbirds.*

pered in their flight by carrying the extra weight of undigested food. This can be a handicap for birds that take larger meals and digest food slowly. Scavengers such as vultures have to eat as much as they can when food is available, and a well-fed vulture is so heavy that it cannot take off at all. A fully "tanked-up" hummingbird, on the other hand, is still hovering.

It is nonetheless still important for hummingbirds to practice economies and reduce their food requirements. This involves quite small adjustments in behavior such as perching on a flower whenever possible to save the effort of hovering. In the long term, hummingbirds save energy by reducing heat loss. As the smallest warm-blooded animals, they must guard against heat loss, because a small body has a relatively large surface area and loses heat rapidly. It is sometimes said that hummingbirds have the fewest feathers of any bird. This is simply because a small body has a small area of skin and requires fewer feathers to cover it. A more interesting fact is that hummingbirds have the highest density of feathers of any bird. A ruby-throated hummingbird has about 1,000 feathers and a brown thrasher nearer 2,000, but the thrasher has 10 times the area of skin, so the hummingbird has five times the density of feathers. Hummingbirds need the dense pile of feathers as insulation to keep warm, especially at the high altitudes of mountain habitats and the high latitudes of the northern and southern limits of the Americas. Good insulation is also necessary in deserts, where keeping warm

at night is a problem for small animals because heat is radiated into the clear night air.

TORPOR

In cold weather, when a bird on a perch would be shivering to keep warm, a hummingbird in flight beats its wings more rapidly to increase the heat generated in the muscles and to save energy overall. Of course, this is of no use at night when the bird is roosting. Hummingbirds solve the problem of cold nights by going into a state of torpor, called "noctivation" by Alexander Skutch in allusion to the similar and more familiar state of hibernation, which can be thought of as prolonged torpor.

When John Gould, famous for his stupendous illustrated book on hummingbirds, was returning to England in 1857 with some live hummingbirds on board, his ship took a northerly course over the Grand Banks near Newfoundland. The hummingbirds responded by becoming cold, stiff and to all appearances dead, but Gould found that they could be restored to full health when placed by a fire or kept inside his jacket. Normally, this torpor occurs only at night, when the hummingbird cannot see to feed. The bird appears almost lifeless, head drawn in and bill pointing up, and it barely moves when handled. It saves energy by reducing its body temperature in much the same way a householder saves on fuel bills by turning down the thermostat at night. Less fuel is required to maintain the house at 65 degrees F than

A rufous hummingbird (Selasphorus rufus) *at rest. During their annual migrations, rufous hummingbirds may use torpor as a means of conserving energy.*

at 70. In normal sleep, a bird drops its temperature by several degrees, but hummingbirds can lower their temperature from over 100 degrees to about 60. Most maintain this level if the temperature of the surrounding air drops further, but the body temperature of the Andean hillstar (*Oreotrochilus estella*) will fall to 40 to 45 degrees if the air temperature goes below freezing. A reduction to about 60 degrees is enough for an 85 percent energy saving. As the body cools, the heartbeat drops to 50 or fewer beats per minute, compared with its normal resting rate of 250, and breathing becomes so irregular that it may stop for several minutes.

Whether a hummingbird goes into torpor depends to some extent on how well it has fed during the day. With plenty of food reserves, it can sleep normally with only a small drop in temperature. If it becomes torpid, its temperature goes down rapidly, and it will raise its feathers to encourage heat loss. Recovery can be equally quick, especially when the morning sun strikes the bird in its roost or it is cupped in human hands. To watch an apparent corpse coming to life and reverting to its energetic life of racing hither and thither is an amazing experience. It is not surprising that Jesuits working in Mexico used the torpor of hummingbirds as a parable to explain the Resurrection of Jesus Christ.

FORWARD FLIGHT

While hovering attracts the most attention, it should not be forgotten that hummingbirds also move in straight and level flight. Here again, their speed of movement has been exaggerated—small animals look as though they are moving faster than they are. Unless radar can be used for tracking flight in a straight line, animal speeds are notoriously difficult to measure except with captive or trained animals. The added complication for flying animals is the effect of wind. A bird flying at 30 mph with a 30 mph tailwind will be recorded as traveling over land at 60 mph.

Pioneering records of hummingbird flight have come from experiments in a wind tunnel carried out by Crawford Greenewalt. He tempted hummingbirds into a miniature wind tunnel with a sugar-water feeder and increased the speed of the fan until the birds were just able to move forward to the feeder. Ruby-throated hummingbirds achieved speeds in the wind tunnel just below 30 mph. This appears to be the maximum for straight and level flight under normal circumstances, but they could probably have flown faster if they had considered it essential to reach the feeder. Faster speeds have been recorded for rubythroats under different circumstances: up to 50 mph when escaping from a rival and 63 mph when a male dives in courtship display. This record has been exceeded by a green violetear (*Colibri thalassinus*) that clocked over 90 mph in short chases.

To fly faster, a bird increases the amplitude (sweep) of its wing beats—the wings travel through a larger arc in the same amount of time. (When they are working hard, hummingbird wings sometimes meet or even cross at the bottom of the downstroke.) They can also override their natural frequency of oscillation—a broad-tailed hummingbird (*Selasphorus platycercus*) increases its wing-beat frequency from its normal 45 bps to 51 bps when chased.

In this respect, a hummingbird is rather like a ship that has an economical cruising speed determined by the shape of its hull. It can make a few extra knots, but only by burning an excessive amount of fuel. Birds are faced with a similar problem. Each species has a cruising speed that requires the least power from the flight muscles and depends on the shape of the wings and the streamlining of the body. As the bird flies faster, the drag on its wings and body increases, and it needs more power to overcome the drag and propel itself through the air. Faster than economical speeds are useful if they save the hummingbird time when it moves between flowers or if it is escaping from a predator. Likewise, male hummingbirds spare no effort in their courtship displays. The spectacle of a male's zooming aerobatics is a distinct attraction for the female, and she may even select a mate on the basis of his flying ability.

MANEUVERS

Hummingbirds' hovering flight also permits immense maneuverability. As well as being able to keep their position in front of a flower nodding in the wind, hummingbirds indulge in spectacular aerobatics during territorial chases and courtship displays. Even in "normal" flight, they weave

A band-tailed barbthroat (Threnetes ruckeri) *in profile shows off its tail feathers and slightly curved bill.*

through gaps in the foliage with ease.

Other birds, too, have a high level of skill—a chickadee can land upside down on a feeder and a hawk chase through a narrow gap by folding its wings with millisecond timing—but high-speed photography has demonstrated that hummingbirds are in a class of their own. Crawford Greenewalt's films showed that when a hummingbird is surprised while feeding at a flower, it can spin around and fly away at full speed in about two-tenths of a second. Startled humans would hardly have begun to react in the same brief time. The hummingbird's response is to remove its bill from the flower with a forward sweep of its spread tail, put its wings into reverse and flip onto its back. Then it does a half-roll so that it is right way up again, facing away from the flower, and ready to streak off.

To carry out such maneuvers at speed requires little more than subtle changes in the attitude of the wings. By tilting the wings in the required direction, as a helicopter pilot tilts the rotors, a forward, sideways or backward component is added to the main vertical lifting force. As we have seen, when a hummingbird is hovering at a flower, its body is tilted upright so that its wings sweep to and fro in a horizontal plane and the lifting force is directed vertically. To fly forward, the body tilts to bring the sweep of the wings toward the vertical. The nearer to the vertical, the faster is the forward motion. The pattern of the wing beat also changes, and the wings no longer turn over on the upstroke. At "full speed ahead," lift is still produced on both up- and downstrokes as in hovering, but forward thrust comes only from the downstroke.

Earlier in the chapter, I stated that the hovering flight of hummingbirds was unique among birds. The qualification is necessary because some insects and bats have lifestyles that make use of sustained hovering. Large day-flying moths such as the hummingbird clearwing moth and other sphinx moths that sip nectar while hovering in front of flowers are often mistaken for hummingbirds, which results in mistaken reports of "flocks of baby hummingbirds." Even in Europe, the hummingbird hawkmoth sometimes deceives people into thinking they have actually seen a hummingbird. Long-tongued bats are the mammalian equivalent of hummingbirds. They live in the warmer parts of the Americas, from northern Mexico to Argentina, and hover in front of flowers to probe for nectar with their long tongues. Although hummingbirds are famous for their hovering prowess, insects and bats are more efficient, and one bat species has been found to require substantially less power for hovering than do hummingbirds.

In view of the high fuel consumption of hovering, an obvious question is why hummingbirds have adopted such an expensive lifestyle and hover at flowers instead of saving energy by perching, as do other nectar-drinking birds such as Australian honeyeaters. One answer seems to be that as a means of discouraging insects and birds other than hummingbirds, hummingbird flowers are not equipped with perches. Another possibility is that hummingbirds can move rapidly between flowers if they do not land and thus can feed much more quickly. This only works if there is plenty of nectar in the flowers. If other birds are competing for nectar at the same flowers, the volume is reduced, and the hummingbirds have to visit more flowers.

FEATHER MAINTENANCE

Like any bird, a hummingbird must keep its plumage in good condition. In regular preening, the bird shuffles the feathers into place, oils them and removes parasites. Birds with damaged, misshapen bills are known to carry larger populations of parasites because they cannot preen effectively, yet hummingbirds with long, unwieldy bills do not suffer any more from parasites than birds with short bills. Some long-billed hummingbirds reach the body feathers by stretching their necks as far as possible and reaching down with their bills.

But the main solution is to "scratch-preen," a leisurely movement compared with ordinary scratching that consists of combing the feathers with the claws. Sometimes scratch-preening is performed on the wing, the hummingbird man-

A violet-headed hummingbird (Klais guimeti) *uses its bill for preening, a vital part of healthy feather maintenance.*

aging to bring a leg over the wing and scratch its neck while flying. Scratch-preening is used to preen the head and chin, but long-billed species also apply it to the body. The sword-billed hummingbird's four-inch bill is so enormously long that it can only reach its wings, and that with difficulty, but the species has unusually flexible legs and feet and can scratch the middle of its back with its claws.

Bathing is a common activity among hummingbirds. They may stand in shallow water and bob and splash like other birds, but they also perch near spraying water, fly through small waterfalls and lawn sprinklers, dive into pools and rub against wet leaves. Or they may simply expose themselves on a perch in the rain, spreading their wings and tail to catch the drops. The function of bathing is something of a mystery and does not seem to be to keep the plumage clean. It is probably to soften the feathers and make them easier to preen.

Eventually, the feathers wear out and must be replaced by the molt. Like most birds, hummingbirds grow a new suit of feathers every year. This is a difficult time for the bird, because it has to find protein for making the feathers as well as extra energy to fuel the manufacturing process. In the meantime, the loss of old feathers makes flying more strenuous. A hummingbird can fly with 30 percent of its wing area missing; it compensates by losing weight to reduce the effort of flying.

Nevertheless, hummingbirds take four to five months to complete the molt process, so their flight is never seriously impaired. Their molt has one unique feature. The primary flight feathers that make up most of the wing surface are molted in sequence, from the innermost to the outermost, as they are in songbirds and certain other groups. But in hummingbirds, the shedding of the two outermost feathers is reversed. This is an adaptation to ensure that the aerodynamic stability of the wing is not impaired while they are hovering. As a rule, birds organize their molt to avoid times of the year when they are facing other demands such as nesting and migrating. Thus North American hummingbirds molt after migrating to their winter homes. Tropical species avoid molting while breeding, but long-tailed and little hermits may molt slowly throughout the year, and females interrupt the molt while nesting.

A crowned woodnymph (Thalurania furcata) *stirs up a storm in a pool of rainforest water.*

HUMMINGBIRDS AND FLOWERS

Chapter Three

When the Europeans who first explored the New World reported seeing tiny jewel-like birds, they enhanced the heavenly image of these exquisite beings by describing how they fed on nectar—the drink of the gods. John Winthrop, Governor of Connecticut, wrote to the English naturalist Francis Willughby in 1670: "'Tis an exceeding little bird, and seen only in Summer...flying from flower to flower, sucking with its long bill a sweet substance." But some confusion endured about just what transpired between the bird and the flowers. Thomas Morton had written in 1632 that the ruby-throated hummingbird was "a curious bird...that out of question lives upon the bee, which hee eateth and catcheth amongst flowers....Flowers he cannot feed upon by reason of his sharp bill which is like the poynt of a Spanish

*Above left, a green hermit (*Phaethornis guy) *and a blue-throated hummingbird (*Lampornis clemenciae)*, right, both demonstrate precision flying as they collect nectar from flowers.*

An emerald-bellied puffleg (Eriocnemis alinae) *supports itself with its feet as it immerses its bill in a flower.*

*A Peruvian white-bearded hermit (*Phaethornis hispidus) *shows off its considerable bill and tongue.*

needle but shorter." That contradictory reports about the feeding habits of hummingbirds filtered back to Europe was not surprising. We now know that although hummingbirds are nectar drinkers, they must supplement their diet with insects.

NECTAR DRINKERS

Watch a hummingbird in good light as it hovers in front of a flower, and you can often see the long, slender, almost translucent tongue emerge from the tip of the bill and flicker in anticipation. The hummingbird looks like a huge butterfly or moth, a comparison which led to the idea that a

hummingbird's tongue is a fine tube for sucking up nectar. In fact, hummingbirds lick nectar; they don't suck it. The tip of the tongue is forked, and each half of the fork curls into a hollow trough. Nectar is drawn into the troughs by capillary action (the force that draws fluid a short distance up a narrow tube when the tip is immersed) and is squeezed off when the hummingbird extends its tongue again for the next lick. The bird flicks its tongue in and out many times per second, bringing a pulsing stream of nectar into its mouth.

The hummingbird's unique way of life depends on this monotonous but bountiful source of food, which it

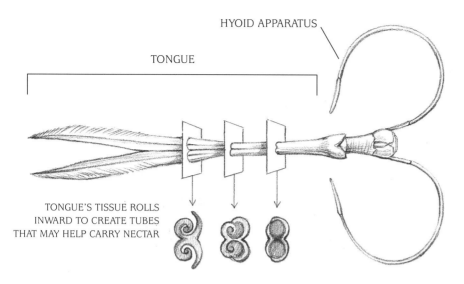

HYOID APPARATUS

TONGUE

TONGUE'S TISSUE ROLLS
INWARD TO CREATE TUBES
THAT MAY HELP CARRY NECTAR

*The action of the hyoid apparatus forces
the hummingbird's tongue beyond its
bill. The forked tip picks up nectar, but
the tongue is not designed for catching
insects as was once believed.*

consumes in large quantities every
day. As described in the preceding
chapter, their energy-demanding
hovering allows the birds to access
energy-rich nectar. The hovering and
the nectar diet are inseparable. The
physical adaptations for hovering are
necessary for hummingbirds to sustain themselves with nectar, and
their entire way of life centers
around the need to obtain sufficient
fuel for hovering.

The wide range of flower types
growing in the Tropics and the birds'
specializations for exploiting their
nectar are the main reasons for the
evolution of such a large number of
hummingbird species. There has
been a coevolution of flowering
plants and hummingbirds: The
plants encourage birds to deliver
pollen from flower to flower by lur-
ing them with the promise of nectar,

and the birds have taken full advantage of the offer by evolving their
unique lifestyle.

Plants with flowers adapted for
pollination by hummingbirds are described as ornithophilous, meaning
bird-loving. Even in North America,
which is the home of only a small
minority of the 318 hummingbird
species, more than 130 species of ornithophilous plants exist. Many are
visited by other nectar-drinking birds
such as orioles. Ornithophilous
plants share a number of features.
The most important is that they have
abundant nectar. Hummingbirds
need large amounts of energy compared with flower-visiting insects,
and they quickly learn to ignore
flowers that cannot be easily drained
of a plentiful supply.

Copious nectar makes a flower
worth visiting, but the flower must

A bearded helmetcrest (Oxypogon guerinii) *perches briefly as it inspects a flower.*

also advertise its nectar and make it easy for the hummingbird to extract. Bird flowers, unlike insect flowers, usually have little or no scent, which is an important attractant for insects. This is not surprising: Most birds have a poor sense of smell but very good color vision. Hummingbird flowers are most often red—the color to which birds' eyes are very sensitive—and less often orange and yellow. It was once thought that hummingbirds had an instinctive preference for red, but now it seems they learn to associate red with nectar.

Researchers have discovered that captive hummingbirds prefer the red flowers of scarlet gilia (*Ipomopsis aggregata*) to the white flowers of the closely related *I. tenuituba*. Presumably, the hummingbirds learned that red flowers were more rewarding. However, the birds will change their preference and select white flowers if they are rewarded with extra nectar. An experiment with feeders filled with variously colored sugar solutions— red, green, blue, yellow and clear— showed that ruby-throated hummingbirds have no particular color preference initially, but once they

find a rich source of sugar, the color becomes a means of finding more of the same. This would explain why the white-eared hummingbird (*Hylocharis leucotis*) has an unusual preference for the blue flowers of *Salvia*. Helmuth Wagner found that when his local birds were feeding on blue flowers, they would choose to visit blue feeders, but when they were feeding on purple flowers, they visited purple feeders.

A note of caution is necessary in any discussion of hummingbirds and flower colors, because it is now known that birds' color vision is better than ours: They see all the colors we do and ultraviolet as well, which is invisible, or black, to us. Some flowers develop ultraviolet patterns when they are most fertile and secreting the most nectar, which may be a way of making themselves even more attractive to hummingbirds.

There are, nonetheless, inherent advantages to red flowers. Red is a color birds can see easily, especially against a background of green. Red flowers are not common in the northern parts of the ruby-throated hummingbird's range, but they are abundant on its migration route. Easily spotted flowers whose color instantly proclaims them as bird flowers may be important for rubythroats when they land in an unfamiliar place and need to find food quickly. Another

advantage of red is that it is not perceived by insects—they will not be attracted to red flowers and into competition with hummingbirds.

Even when insects are attracted to hummingbird flowers, they are discouraged from taking nectar, because with a few exceptions, they have to land in order to feed. The same is true of other nectar-feeding birds. Flowers attractive to insects typically have petals that form a convenient landing platform, or if the flowers are small, they are clustered into heads that allow the insect to walk from flower to flower. Hummingbirds would probably prefer to land on flowers to avoid the cost of hovering, but hovering enables them to feed at flowers that typically hang downward or have short or curling petals and thus do not provide a platform for insects or other birds.

Insects and birds other than hummingbirds sometimes steal nectar from hummingbird flowers by biting holes in the base of the flowers. This shortcut strategy allows them to cheat the plant by taking the nectar while bypassing the stamens and avoiding picking up pollen. When bumblebees rob scarlet gilia flowers, the flowers are visited significantly less often by hummingbirds, which quickly learn they are not worth visiting. As a result, the flowers set fewer seeds. Many tropical

The white-tipped sicklebill (Eutoxeres aquila) *specializes on curved flowers such as the Costa Rican* Heliconia reticulata.

hummingbird flowers have another way of reducing the effects of cheating. They open for only one day and secrete their nectar before dawn, when they are visited by hummingbirds, which are active at first light, before bees and other cold-blooded insects have had time to warm up and set out in search of food.

The sin of cheating is compounded by some hummingbirds, which either use existing holes made by the insects or, in the case of some species with short, sharp bills, make their own. The rufous-tailed hummingbird (*Amazilia tzacatl*) of Colombia takes advantage of holes made by another bird, the bananaquit. The ruby-throated hummingbird cheats cardinal flowers by feeding from holes in the base of the corolla at the beginning of their flowering season. Later, they feed from the front of the flower and are important pollinators.

The story of competition between hummingbirds and insects for plant resources is not complete without mention of another group of animals. Certain mites that live in flowers, drink their nectar and eat their pollen compete with hummingbirds for food. But they also rely on them for transport. The mites lay eggs and pass through their life cycle in one flower but must hitch a ride on a hummingbird to spread to other

flowers. As a hummingbird's bill enters a blossom, the mites quickly jump on and take refuge in the bird's nostrils. The mites can recognize the nectar of their "own" species of plant, and when the hummingbird visits the right flower, they hop off and settle into their new home.

FINDING FLOWERS

Part of the popularity of hummingbirds lies in their tame and confiding nature. Everyone delights in the tiny birds, and many people feel a special empathy with the ones that fearlessly visit their yards and feeders and even hover around their admiring human hosts. But hummingbirds' reasons for being so trusting are prosaic. Because stands of flowers are continually changing as different species bloom and fade, hummingbirds are continually searching for new supplies of food. They must therefore be curious and examine all likely sources. If they are accustomed to feeding at red flowers, they will investigate anything red, which leads to the inspection of red tin cans, signs, clothing and even red lips. They approach people in their yards or campsites and even adopt and follow them around. There is a story of a man who was "befriended" by a rufous humming-

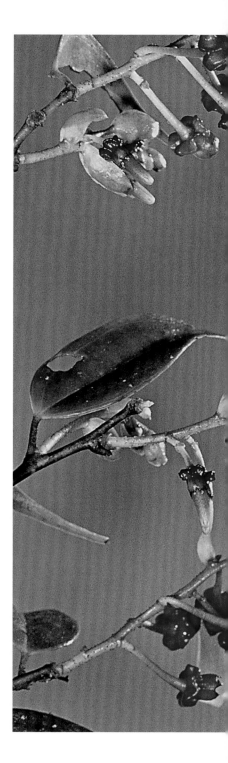

The female purple-throated mountain gem hummingbird (Lampornis calolaema) *is a trapliner, while the male is a pugnacious defender of a territory.*

bird that followed him on daily walks and eventually hitched a lift by perching on him. When he returned home after a month away, the hummingbird was there to greet him, hovering around his head.

Hummingbirds must also have good memories and save time and energy by recalling the location of flowers. Rufous hummingbirds, and presumably other species, avoid flowers they have recently visited and drained of nectar. Migrants even remember where they have found food on one stopover and return to search for it when they pass by a year later. There are stories of hummingbirds appearing in spring and hovering at the exact spot where a feeder had been hanging the previous year.

INSECT SUPPLEMENTS

Popular accounts of hummingbirds focus on their habit of feeding on nectar, not only because it is the trait that brings them into close contact with people but also because the evolution of hummingbirds has been driven by their relationship with flowers. Yet it has long been known that hummingbirds also feed on insects and spiders. Some early naturalists, including Charles Darwin, believed that hummingbirds ate nothing but these animals, because specimens shot for museum collections had only tiny insects and spiders in their stomachs. They decided that visits to flowers by hummingbirds were for the purpose of collecting insects, not realizing that nectar passes through the digestive system

so rapidly that it leaves little trace when a hummingbird is dissected.

The amount of animal food taken varies among hummingbird species, but insects are important for many hummingbirds when flowers are scarce. Rubythroats have been observed defending rotting fruit under trees that has attracted insects. In the forests of Costa Rica, where there is a marked dry season, hummingbirds eat mainly insects from January to March and revert to nectar-drinking when the plants bloom after the start of the rains. Similarly, before the spring burst of flowering in North America, four species of hummingbird (ruby-throated, Anna's, rufous and broad-tailed) visit holes that sapsuckers have drilled in trees. Here, the birds both drink the sap and catch other insects attracted to the holes. For the ruby-throated hummingbird, sap is so important that the northern limit of its spread up the east side of the continent is probably determined by the range of the yellow-bellied sapsucker. However, ornithologists have pointed out that a diet of insects cannot provide enough energy for hummingbirds' expensive lifestyle.

Hummingbirds need as much protein in their diet as any animal. Nectar provides sugars in abundance but only minute quantities of the proteins, salts and other substances required to sustain life. The supplement of insects and spiders is needed to remedy these deficiencies. Captive hummingbirds are known to increase the number of insects caught before and during the molt, when they need extra protein for

The long bill of the green-fronted lancebill (Doryfera ludovicae) *gives it an advantage at tubular flowers.*

manufacturing feathers. If no insects are available, captive hummingbirds have been observed chasing dust particles illuminated by sunlight and even completely imaginary prey, then going into a frenzy of insect-hunting when insects are introduced.

For well over a century, ornithologists had quite the wrong idea about the way hummingbirds catch insects. In 1861, John Gould described how "short-billed Lesbiae cling to the upper portion of those flowers, pierce their bases, and with the delicate feelers at the extremities of the tongue, readily secure the insects which there abound." A hundred years later, researchers investigating the anatomical structure of the hummingbird tongue still believed that "insects are entangled in the fimbriated [having a fringed border] tip of the tongue when the hummingbird is probing into flowers or picking insects from some surface." They were all wrong. The hummingbird tongue does not entangle insects. And unlike a chameleon's tongue, it is not sticky, which has also been suggested. On the contrary, if an insect accidentally sticks to its tongue, a hummingbird will shake its head, flick its tongue or scratch its bill until the insect is dislodged. Walter Scheithauer went so far as to press tiny flies onto the tongues of his captive hummingbirds and found that they always fell off.

If its tongue is not used for catching insects, the problem becomes one of how the bird gets an insect to the back of its long bill then works it into the mouth, where it can be swallowed. The hummingbird meets the challenge in one of two ways, depending upon whether it is hawking for flying insects or gleaning them from leaves. When hawking, by either flying from a perch as a flycatcher does or darting around a swarm of insects, the hummingbird flies with bill agape so that the insects go straight into the back of the mouth. The same technique is used for taking spiders (and their trussed-up prey) from their webs. The hummingbird compensates for the narrowness of its bill and gape with extreme maneuverability, which allows it to strike with great accuracy (hawkers are typically species with short, straight bills). When gleaning, a hummingbird flies around leaves and tree trunks and plucks its prey by grasping it with the tip of the bill, tossing it into the air, then flying forward to take it into its mouth.

Observations of captive hummingbirds have revealed that, in at least some species, the bird disturbs the insect with the tip of its open bill and sweeps it back into the mouth with vortices of air created by the beating wings, which form a "funnel"

A stripe-tailed hummingbird (Eupherusa eximia) *steals nectar from near the base of an* Aphelandra *flower in the Costa Rican cloud forest.*

along the length of the bill. A species may employ either tactic, but the curve-billed hermits, which are closest to the ancestral insect-eating hummingbirds, almost always glean by hovering under leaves.

A.J. Mobbs has witnessed two variations of the gleaning technique in aviary birds. A blue-chinned sapphire (*Chlorestes notatus*) hovered over a resting insect, opened its bill, advanced and tapped the leaf sharply with the tips, making the insect fly up into its mouth. A violet-headed hummingbird (*Klais guimeti*) seized an insect but, rather than tossing it into the air, manipulated its upper and lower bills to work the insect into its mouth. In the cool heights of the Andes, where insects are likely to be found crawling over the meadows and the thin air makes hovering difficult, helmetcrests (*Oxypogon*) and thornbills (*Chalcostigma*) find it more profitable to walk in search of prey on legs that are unusually strong for hummingbirds.

BIRD-FLOWER RELATIONSHIPS
Naturalists watching hummingbirds soon came to realize that there is often a correlation between the length

The snowcap hummingbird (Microchera albocoronata) *is a tiny species that feeds on flowers frequented by insects.*

of a hummingbird's bill and the length of the corolla tubes of the flowers it visits regularly. Charles Darwin was one of the first to be fascinated by the match between flowers and the animals that pollinated them. He concluded that the relationship between a plant and its pollinator had a profound effect on the evolution of the incredibly diverse array of flower shapes and colors around the world. He pointed out that the "beaks of hummingbirds are specially adapted to the various kinds of flowers they visit."

Bill length ranges from ⅜ inch in the purple-backed thornbill to that of the sword-billed hummingbird, whose enormous four-inch bill is as long as the rest of the bird. Swordbills in Colombia are dependent on a single species of passionflower, *Passiflora mixta*, whose corollas are about 4½ inches long. The hummingbird makes up for the difference in length by extending its tongue to reach the nectar at the base of the corolla. No other hummingbird can reach the nectar of this passionflower, so the swordbill enjoys an exclusive food supply.

The other really remarkable bill is that of the sicklebills (*Eutoxeres*), which is as curved as the name suggests. John Gould wrote, "It is evident that its singularly shaped bill is adapted for a special purpose, and we may readily infer that it has been expressly formed to enable the bird to obtain its food from the deep and remarkably shaped flowers...for exploring which a bill of any other form would be useless." In fact, the white-tipped sicklebill (*E. aquila*) feeds exclusively on the curved flowers of *Heliconia pogonantha*, a wild plantain, but unfortunately for Gould's hypothesis, the flowers are visited by a variety of other hummingbird species, including some with straight bills.

The correlation between the lengths of hummingbird bills and flower corolla tubes is often given as a good example of coevolution, in which two species evolve characteristics for their mutual benefit. The idea is that a hummingbird feeds most efficiently at flowers that match the length of its bill, because the quantity picked up is inversely proportional to the distance the tongue protrudes beyond the tip. If the bird has to stretch its tongue to reach the bottom of the flower, it does not pick up as much nectar. So,

the theory goes, a short-billed hummingbird feeds more efficiently at short flowers, where it does not have to extend its tongue far, while a long-billed hummingbird prefers to visit flowers with long corolla tubes to avoid competition with species that have shorter bills.

Flowering plants appeared on Earth at the end of the Mesozoic era, roughly 70 to 100 million years ago, and rapidly evolved in conjunction with flying insects. The plants developed showy, colorful flowers with scents and nectar to entice the insects to visit them and transfer pollen from flower to flower. Millions of years later, hummingbirds appeared on the scene. At first, they were insect eaters with slender bills for snatching rich pickings from the exuberant year-round crop of flowers in the South American Tropics. Gradually, they turned to stealing the insects' food by sipping nectar, although they still continued, to a greater or lesser degree, to catch insects. At the same time, the plants began to take advantage of the hummingbirds in the same way they had the insects, by adapting their flowers specifically to attract hummingbirds as pollinators.

Coevolution with flowers is one aspect of hummingbird life that interests biologists. The evolution of

bird-pollinated plants from insect-pollinated plants in North America has been investigated by Karen and Verne Grant, who have shown that hummingbird flowers are most likely to have developed from those pollinated by bees. Hummingbirds still visit a wide variety of open, tubular bee flowers that provide easy access to long tongues. Gradually, through the process of natural selection, some bee flowers evolved characteristics that are less attractive to insects and more useful to hummingbirds, and at the same time, they modified their pollination mechanisms to favor transport of pollen by hummingbirds.

As evidence, the Grants cite several types of North American plants that are mainly bee-pollinated but which have a few species modified for pollination by hummingbirds. Of about 50 species of larkspur (*Delphinium*), almost all are bee-pollinated, but two West Coast species (scarlet and canon delphinium) have bird flowers. The Jacob's ladder genus (*Polemonium*) includes the bird-pollinated *P. pauciflorum* in southern Arizona. Sages (*Salvia*), locoweeds (*Astragalus*), currants (*Ribes*) and honeysuckles (*Lonicera*) are also mainly bee-pollinated but have a few hummingbird-pollinated species. Most of the penstemons are

Pollen sticks firmly to the bill of this magenta-throated hummingbird (Philodice bryantae) *until it is deposited in another flower.*

pollinated by bees and have blue flowers, but *Penstemon centranthifolius* is bird-pollinated and has red flowers. The columbines (*Aquilegia*) are a group of plants that originated in the Old World, where there are no hummingbirds, and spread to the New World, where they have adapted to bird-pollination. Not all hummingbird flowers are derived from bee flowers; a few North American bird flowers in the catchfly (*Silene*) and lily (*Lilium*) genera have ancestors that were pollinated by butterflies and moths.

The relationship between the swordbill and its passionflower is unusually close. Among North American species, a good "fit" can be seen between the ruby-throated hummingbird and basal balm (*Monarda clinopodia*), the rufous hummingbird and *Penstemon labrosus* and the calliope hummingbird and scarlet gilia, but none of these is an exclusive relationship. The old idea of an exclusive relationship between one hummingbird and one flower does not exist in reality, and the neat story of hummingbirds visiting the flowers that fit their bills is spoiled by their frequent visits to flowers that may be considerably

longer or shorter than their bills. Moreover, hummingbirds are not restricted to feeding at bird flowers. During the hummingbird breeding season, the rufous feeds mainly on insect-pollinated flowers such as blueberry and salmonberry. Only toward the end of the summer do hummingbird-pollinated flowers become plentiful.

Some experiments with artificial flowers suggest that although long-billed hummingbirds have better success at long flowers than do short-billed hummingbirds, the latter have no advantage at short flowers. It would appear, then, that all hummingbirds ought to have long bills. But there is a disadvantage to long bills. Long-billed hummingbirds have difficulty aiming at narrow flowers. It takes them longer to line up the tip of their bill with the opening, and if they miss, they have to back off and try again. They can even become trapped in narrow flowers; ruby-throats have been seen retiring to a perch to remove the corollas of basal balm caught on their bills.

It seems that the width of the corolla's opening may be as important as its length. Wider corollas allow hummingbirds to extract nectar

One wing of this broad-billed humming-
bird (Cynanthus latirostris) *is splayed*
so that the bird can maneuver from one
source of nectar to the next.

more rapidly, because they can reach deeper into the flower and the tongue has less distance to travel. Rubythroats can insert their bills only halfway into an artificial flower $\frac{1}{16}$ inch wide. When the flower is $\frac{1}{4}$ inch in diameter, they push their faces into the corolla to get closer to the nectar, thus reducing the time needed to drink.

In any plant community, it is normal for a variety of humming-birds to visit a variety of flowers. An overly specialized plant-humming-bird relationship has disadvantages. Any catastrophe that significantly depletes the population of the plant or the bird would have serious con-sequences for the other. If the plant disappeared, the bird would go hungry; if the bird became rare, the plant would fail to set seed. Such a disruption in the ecosystem is of concern to conservationists (see Chapter 7).

POLLEN TRANSFER

But why should plants want their flowers to match hummingbird bills? The payoff for the plants is that by discouraging some visitors and en-couraging others to specialize in vis-iting their species, their pollen is

more likely to be transferred from one plant to another and effect cross-pollination. The main strategy for bird-pollinated plants is to discourage long-tongued bees and butterflies, which are hummingbirds' competitors for nectar (night-flying moths are excluded because bird flowers open in the daytime). The nectar of flowers with long corollas is often beyond the reach of bees and butterflies, and bird flowers are, in any case, not attractive to insects because of their color and their lack of scent.

Most hummingbird flowers have stamens and styles that protrude from the corolla and rub against the bird's feathers or bill. Hummingbird flowers are typically large and grow singly or in loose clusters from flexible stalks. The possible reason for this pliability may be to make it more difficult for the hummingbird to feed. When rubythroats feed at spotted jewelweed, they spend more time at free-swinging flowers than at ones that are held immobile experimentally. The extra time spent at the free-swinging flower increases the amount of pollen they pick up. This may also be the reason why many hummingbird flowers have a plentiful supply of nectar. With scarlet gilia, the amount of pollen picked up from the stamens and deposited on the stigma by a visiting hummingbird depends on the time the bird spends probing for nectar. Of course, this is not in the best interests of the hummingbird, which needs to spend as little time as possible at a flower. The coevolution of plants and hummingbirds is not just

After piercing the base with its bill, a stripe-tailed hummingbird (Eupherusa eximia) *steals nectar from a flower in the Costa Rican cloud forest.*

Some bats, such as the Geoffroy's tailless bat (Andura geoffroyi) *seen here, also hover at flowers to feed.*

about cooperation; it is also about conflict of interest.

We have seen how North American hummingbirds will visit any suitable flower to collect nectar—even ones that do not supply any, such as the cardinal flower. To be effective as pollinators and so justify the plants' investment in nectar production, the hummingbird must transfer pollen between flowers of the same species. An individual hummingbird may carry pollen from three or four different species on its body at any one time, and flowers frequently receive the wrong type of pollen. But this does not matter as long as some of the pollen gets to the right destination.

Details of how plants can improve the chances of cross-pollination were investigated by James Brown and Astrid Kodric-Brown in the White Mountains of central Arizona. They found nine species of hummingbird flowers in one area. All had the general features of red color, long corolla and plentiful nectar, so there was no signal that would lead hummingbirds to specialize in visiting a single kind of plant. Instead, the plants attempt to increase the chances of exchanging pollen with their own species by depositing it only on a particular spot on the bird's body. When the bird visits another plant of the same species, the stigma is positioned to pick up

Unable to hover, the marico sunbird (Nectarinia mariquensis), *above, instead sips nectar while perched. A nectar and pollen collector, the honeybee, right, prefers open flowers.*

DAILY AND SOCIAL LIVES

Chapter Four

It is easy to get the impression that hummingbirds lead a continuously busy life. At a bed of flowers or a sugar feeder, there is a kaleidoscope of hummingbirds buzzing to and fro, sipping nectar and then zooming off into the distance. Oliver Pearson observed a very different picture, however, when he spent two days watching an Anna's hummingbird in the botanic garden of the University of California at Berkeley. This hum-mingbird was sustained by 1,022 flowers growing inside its 50-foot-diameter territory. It fed for only two hours of the day and spent 10½ hours (80 percent of its active hours) perching. Apart from five minutes of chasing insects, the hummingbird passed the balance of its feeding time hovering at flowers. Other activities included about 15 minutes of territory defense and moving from perch to perch.

Above left, a calliope hummingbird (Stellula calliope), the smallest bird in North America, visits a backyard feeder, while a long-tailed sylph (Aglaiocercus kingi) sips nectar from a flower, right.

Given the energy demanded by hovering, it is not surprising that hummingbirds try to spend as much time resting as possible. Their technique is to fill their crop (a throat pouch in which birds store food) and retire to a perch to digest. After about four minutes, the bird has moved the nectar from crop to digestive system and is ready for another bout of feeding.

Active time is greater for hummingbirds that have to cover larger areas in search of food and for those with demands such as breeding, courting and gaining weight for migration. The more flowers they defend, the better their food supply, but they will also have to spend more time and energy defending it for their exclusive use. One would expect hummingbirds to find a happy optimum and defend enough resources for their needs without wasting energy, and this is precisely what they appear to do.

In the fall, when rufous hummingbirds migrate south through the Rocky Mountains from their breeding grounds in western North America to southern Mexico, they travel in a series of hops, stopping off in mountain meadows to feed on Indian paintbrush and red columbine. At this time of year, the weather is changeable, and wind and rain can

The broad-tailed hummingbird (Selasphorus platycercus) *heads south in the fall, feeding in mountain meadows on such plants as Indian paintbrush.*

destroy the flowers, so the birds' best policy is to cover the distance as fast as possible. At each stop, they establish territories in which they can fill up efficiently, then they set off on the next leg quickly.

Clifton Lee Gass found that the number of rufous hummingbirds in a meadow varied from day to day according to the number of flowers in bloom. One day, for example, there might be only 25 flowers and no hummingbirds, but a week later, 3,200 flowers and 15 hummingbirds might be present. Individual hummingbirds would also make daily adjustments to their territories so that they always had enough nectar to sustain themselves. This was not simply a matter of defending a certain number of flowers. The amount of nectar they contained was also important. Red columbine produces four times as much nectar as Indian paintbrush, so a territory with a mass of columbines will be smaller than one in which Indian paintbrush predominates. But hummingbirds do not live in a vacuum. They have to defend their nectar from other hummingbirds and other animals such as bees and butterflies. This takes energy, but the hummingbirds have a cunning strategy for dealing with the problem. To foil their competitors, they start each morning by removing nectar from flowers on the edge of the territory so that intruders cannot sneak in and steal it unnoticed.

To test whether the rufous hummingbirds were defending the optimum number of flowers, Lynn Carpenter and her colleagues attached sensitive balances to the humming-birds' favorite perches. The researchers read the balances through a telescope and recorded the birds' changes in weight with each meal. They found that the birds modified their territories by trial and error until they had adjusted the profit gained by food intake against the loss resulting from territorial defense to get the best weight gain per day. A hummingbird with too small a territory gained weight slowly because it did not have enough flowers, but so did one with too large a territory because it spent too much energy in defense. One of the birds started by defending 1,970 flowers and gained only 0.15 gram. The next day, it defended 3,320 flowers and gained a little more—0.25 gram—but that was still too many flowers to defend properly. On the third day, it halved the number of flowers and achieved a happy medium, increasing its weight gain to 0.35 gram. After another day's feeding at the same rate, it had put on enough weight to continue its migration. The conclusion was that these hummingbirds were indeed "optimally foraging" to maximize their energy gain.

DIVIDING RESOURCES

A defended territory is essential for the survival of many hummingbirds. Unlike many other birds, in which a territory is defended by males seeking mates and often has only a secondary function as a feeding ground, many species of hummingbirds defend territories that ensure them a constant source of nectar. The territories are held by males and to a

The tiny coppery-headed emerald (Elvira cupreiceps) *is a trapliner that often feeds at flowers more typically visited and pollinated by insects.*

lesser extent by females and even juveniles at any time of year, including during migration. Sometimes smaller females cannot compete with their own males and have to feed where flowers are so sparse that they are not worth defending.

Hummingbirds defend territories not only against other hummingbirds of both sexes and other types of birds but sometimes against moths, butterflies and bees, all of which compete for the same nectar. In an amazing story of adaptation among birds, insects and plants, there is an example of a moth that succeeds in escaping the attention of a particular hummingbird species. Green thorntail hummingbirds (*Popelairia conversii*) defend territories on giant Inga trees in the forests of Colombia. They ignore the white-banded sphinxlet moths, because they mimic female hummingbirds in their appearance and behavior, but they do attack tropical kingbirds, which prey on the moths. If male thorntails are excluded from an Inga tree, predation on the sphinx moths rises by 60 percent, which results in a significant improvement in seed production.

For males, territories acquire a second function as sites for attracting females during the breeding season.

Males of the black-chinned hummingbird in flower-rich sites may be densely packed in territories no more than 10 feet across, but the desert-dwelling Costa's hummingbird needs a breeding territory of two to four acres to find enough flowers to sustain itself. If a breeding territory cannot provide all the nectar required, the occupants may fly farther afield to supplement their diet. A Costa's hummingbird nesting in California has a territory centered on a tree and defends the surrounding larkspurs, but some of its foraging takes place outside the territory. If nectar becomes seriously scarce, territories are abandoned and the males search for better flower patches, which are more likely to attract females.

Feeding territories are practical only when flowers are sufficiently abundant in a small enough space for a hummingbird to defend effectively. Typically, territories are set up in flowering bushes and trees or in flower meadows. The alternative is for the hummingbird to forgo defense of a particular territory and forage over a wide area by following a more or less regular route that takes in scattered flowers. This behavior is called "traplining," an allusion to a human hunter making the rounds of his line of traps. Territorial defense

Two Anna's hummingbirds (Calypte anna) *engage in a vigorous midair battle, employing their bills as weaponry.*

and traplining divide hummingbirds into two main ecological groups according to foraging habits. As might be expected from everything else we know about hummingbirds, their place in one category or the other is intimately related to their flight and feeding characteristics.

Territorial species are generalists with medium-length bills that feed on any suitable flowers in their territories. Examples are found among the emeralds, sunangels and pufflegs. Typically, they have short wings, which give them a high wing loading (the ratio of weight to wing area) and a rapid wing-beat rate. Both contribute to the rapid acceleration and maneuverability the birds need for defending a territory. The drawback is that hovering is more expensive for these birds than for trapliners, but an advantage is that the territorials spend less time and energy visiting the tightly clumped flowers to which they have exclusive access.

By contrast, trapliners tend to be specialists with a close relationship between bill shape and their favorite flowers; hermit hummingbirds, with curved bills suited to curved flowers, are a good example. But unspecialized trapliners visit open insect-pollinated flowers as well. Trapliners generally have longer wings and a lower wing loading than territorials. They have less maneuverability but hover more efficiently and spend relatively less energy visiting scattered flowers. Trapliners also fly faster between flowers than do territorials. Frank Gill calculated that a long-tailed hermit can save about two seconds traveling between two flowers 50 yards apart if it flies at 26 mph instead of 18. Two seconds is the time it takes a hummingbird to drain one flower of nectar, and it will gain much more energy from the nectar than it expends by flying faster. Thus flying faster is a cheap way to visit more flowers.

In North America, usually only one species of hummingbird is abundant in any one place, but in heavily forested tropical and subtropical regions with mosaics of patchy habitats are communities of hummingbirds known as "guilds." Twenty or more hummingbird species may coexist in one guild, and they divide up the habitats and the rich and varied arrays of flowers according to their foraging habits. The composition of each guild depends on the types of hummingbirds present and the local environment, which includes both the range of suitable flowers and their flowering times.

Guilds of hummingbirds and other tropical species are studied by ecologists interested in the ways sim-

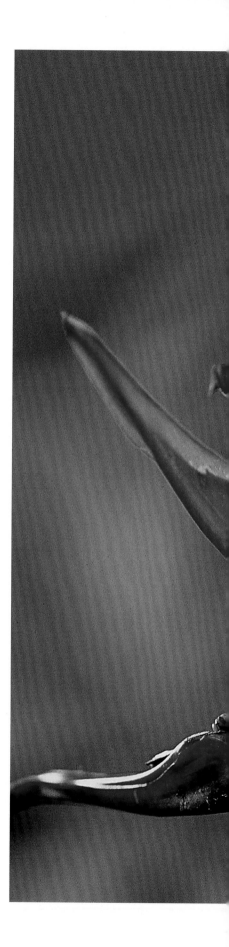

The male fiery-throated hummingbird (Panterpe insignis) *allows the female to feed in its territory, perhaps to increase their offspring's chances of survival.*

Its spread tail provides extra lift as the mountain velvetbreast (Lafresnaya lafresnayi) *hovers before a flower.*

ilar species exploit a common resource. Nectar is a good resource to study because it is easy to measure. It is likewise easy to study the hummingbirds feeding on it and determine the role each type of bird plays in the guild. In Central Mexico, for example, there is a guild of 21 hummingbird species that fall into three broad categories. One category is permanent residents, which are territorials that defend their flowers. The second consists of birds that cannot find places to set up territories. They are nomads and feed by traplining. At certain times of the year, these groups are joined by migrants, which hold territories in places with a surplus of flowers not used by the territorials and nomads. Another category found in some guilds is that of territorial parasites, which are either large enough to ignore the threats of resident territorials or small enough to feed in a corner of a territory without attracting the notice of its owner.

A study of the hummingbirds of Monteverde in Costa Rica shows how flight behavior is related to foraging habits in a community of hummingbirds. Two common hummingbirds, the blue-vented hummingbird (*Amazilia saucerottei*) and the magenta-throated woodstar (*Philodice bryantae*) are territorials that defend dense clumps of flowers with high-speed chases and "dogfights." A large part of their daily energy budget is spent in defense. The fork-tailed emerald (*Chlorostilbon canivetii*) is a trapliner that avoids defended clumps of flowers and flies along a regular route searching for scattered flowers. Occasionally, it meets another member of its own species, and a brief altercation ensues, but nothing as fierce as the combats of the territorials. Calculations show that the fork-tailed emerald uses 35 to 40 percent less energy in hovering than do the two territorial species. For trapliners, life is in the slow lane.

Two more species, the green violetear and the stripe-tailed hummingbird (*Eupherusa eximia*), are territorials or trapliners according to circumstances. They are kept away from defendable clumps of flowers when the more aggressive territorials are present and have to change to traplining. Their flight characteristics are intermediate between those of strict territorials and strict trapliners. Interestingly, in a fifth species, the purple-throated mountain gem (*Lampornis calolaema*), males are territorials and females are trapliners. As may be expected, the males have a higher wing loading and better maneuverability and the females a lower wing loading and more efficient hovering. North American examples of similar differences between the sexes in wing loading and feeding strategy can be seen in Anna's and broad-tailed hummingbirds; in both species, the males tend to be territorials and the females trapliners.

CONFRONTATION AND COURTSHIP

De Lattre's saberwing (*Campylopterus delatterei*) "selects a flowering shrub which it never quits and from which it chases with anger all the

other species of the family that may seem desirous of approaching it." So wrote John Gould. He is not the only person to have been struck by the pugnacity displayed by hummingbirds in defending their food supply. As well as confronting intruders by launching attacks with feet and bill and chasing them from the scene, hummingbirds defend their territories by warning off potential trespassers with eye-catching displays and feeble songs.

Displays take the form of spectacular high-speed dives and other aerobatics, the functions of which are not well understood. The complete sequence of events is often impossible to follow, as the birds race through the air so quickly that it is difficult to trace the circumstances and results of their interactions. Interpretation is even more difficult because the birds use similar displays in both territorial defense and courtship. In hummingbirds, males and sometimes females defend feeding territories and use dive displays to threaten intruders of either sex and often of other species. Males use the same displays to threaten rival males during the breeding season, but they also direct them toward females. This appears to be a preliminary part of courtship, because as in many other kinds of birds, the male is initially aggressive to the female and changes from threatening to courting behavior only when she holds her ground.

Among North American species, the song and displays of Anna's hummingbird are the most impressive and also the best studied in the

The slightly curved bill of the planalto hermit (Phaethornis pretrei) *enables it to feed from elongated rainforest flowers.*

research projects of Gary Stiles of the University of California. The male Anna's hummingbird has one of the more elaborate songs. It has been transcribed as *bzz-bzz-bzz chur-ZWEE dzi! dzi! bzz-bzz-bzz*. Each unit lasts five to six seconds and is repeated one to three times, preceded by a few extra *bzzes*, but it becomes continuous when another male approaches. The dive display of Anna's hummingbird starts with the bird hovering 6 to 12 feet over an intruder—another hummingbird, some other small bird or even a human—and singing several *bzz* notes before climbing almost vertically to 60 to 120 feet. It then dives on the intruder at a breakneck speed of up to 65 mph before pulling up a foot or two above the intruder with a short, loud squeak audible several hundred yards away.

The squeak is produced as the airstream whistles through specially shaped tail feathers. This was cleverly demonstrated by T.L. Rogers, who mounted the feathers on a whip and produced the squeak when he cracked the whip. Some other species produce similar mechanical sounds with modified tail or wing feathers. In male broad-tailed hummingbirds, for example, narrowed feathers at the wingtips produce a whistle; when re-searchers glued the feathers together, the whistle was silenced and the birds became less aggressive and lost their territories.

The dive performance may be repeated 5 to 10 times in rapid succession. The squeak has extra impact when the dive is oriented toward the sun so that the intruder gets a glimpse of a brilliant red flash from the iridescent crown and gorget as the bird passes overhead. The hummingbirds seem to be conscious of the effect, because on cloudy days, the dive is not only performed less often but is also oriented in a different way.

Dive displays are a feature of North American hummingbirds; some perform a U and others a loop. Costa's hummingbird dives from about 100 feet, and the whistle that accompanies the dive has been described as "terrifying in its intensity" and "like the shriek of a glancing bullet." Emily Dickinson's lines on the ruby-throated hummingbird's display are more poetic:

> *A route of Evanescence*
> *With a revolving Wheel—*
> *A Resonance of Emerald—*
> *A rush of Cochineal….*

Gary Stiles has pieced together the courtship sequence of Anna's hummingbird from hundreds of in-

The magenta-throated woodstar
(Philodice bryantae) *has a limited breeding range; where it travels after breeding is unknown.*

complete observations. These were all he could make because the pairs would meet and race away in chases that carried them hundreds of yards from the male's territory. The chases ended with the female perching among thick vegetation and the male singing from a nearby perch or performing the "shuttle display" just above her: flying rapidly to and fro in a short arc of 1 to 2 feet as if he were the weight on a pendulum.

Hummingbirds' shuttle displays are very distinctive and often accompanied by characteristic loud sounds made by the flight feathers. Stiles interprets them as a means of preventing the female from flying farther away. Along with the burst of singing, the displays are likely to be important factors in her decision about whether to accept her suitor. After mating, the female returns to her nest-building, and the male returns to his territory and carries on displaying in the hope of finding another receptive female.

Surprisingly, very little is known about the courtship of many hummingbirds despite their popularity and, in North America at least, their eye-catching courtship performances. Freedom from a parental role has allowed male hummingbirds to concentrate on elaborate courtship displays. This is the reason for the general rule that hummingbird males have the brilliant plumage which has made them so spectacular to humans. They use it for displaying to rivals and attracting mates, while females have drab plumage so that they can sit inconspicuously on the nest.

Study of hummingbird courtship behavior has been complicated partly because of the ambiguous nature of the displays and the difficulty of determining whether they are aggressive or sexual. The problems are greatest in species in which a male courts in the territory where he guards the flowers. If he displays to the female, he may simply be trying to drive her away from his food.

The study of courtship has also been hampered by its brevity and by the challenge of following two tiny birds moving at high speed. All hummingbirds are polygamous; males attempt to mate with as many females as possible, and the bond between them lasts not much longer than is necessary for fertilizing the eggs. The female searches for a mate while she is building her nest or after she has completed it. Attracted by songs or displays, she approaches a male in his territory. He displays to her, first as a rival but, as she stands her ground, increasingly as a mate. Sometimes there is an aerial *pas de deux* as the pair flies side by side.

The eye-catching dive displays seen in North American hummingbirds are repeated in Central and South America among hummingbirds that live in open country. The displays are part of the courtship rituals of the rose-throated hummingbird (*Selasphorus flammula*) of the Costa Rican mountains and of the Andean hillstar of Peru. However, such displays are missing from the repertoire of species living in the depths of forests, where visual displays would be ineffective. Forest hummingbirds make more use of song to attract mates, although their voices are hardly tuneful.

The ruffled head feathers of a Costa's hummingbird (Calypte costae) *may be used as an intimidation tactic during territorial struggles.*

TINY TRAVELERS

Chapter Five

As we know, hummingbirds need a constant supply of nectar-secreting flowers. Once the supply fails, they must move on or perish. The change in venue may entail no more than a short trip. As the flowers on a tree or in a patch of meadow fade, the birds search for a new crop of blooms. In the Chiricahua Mountains of Arizona, for example, the broad-tailed hummingbird exploits the early-flowering western blue flag to fuel the start of its breeding season. As the blooms subside, it moves on to beard tongue, which is just starting to flower. In the Tropics, some species have a daily altitudinal migration: The sparkling violetear, for example, ascends to feed in the open páramo early in the day then retires to the forest below in the afternoon.

Some populations also make complex annual movements be-

Above left, a Florida mangrove, an increasingly popular stop-off point for the migrating rufous hummingbird (Selasphorus rufus), right.

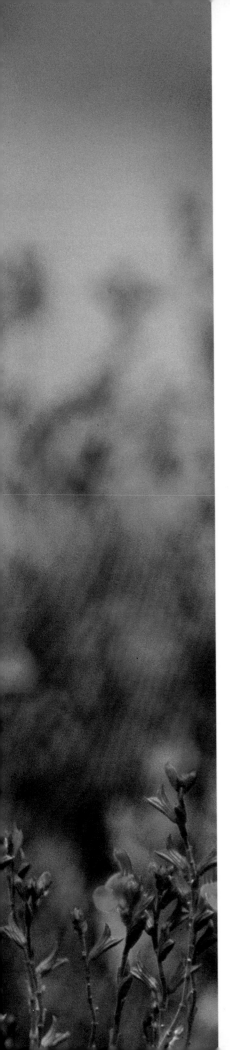

tween different habitats in order to visit a series of favorite flowers. The fork-tailed woodnymph (*Thalurania furcata*) of Costa Rica spends the breeding season in the forest, males occupying the tree canopy and females the understory. After breeding, both sexes move into forest edges, swamps, riverbanks and cleared areas, where they will find nectar-rich *Heliconia* flowers. After feeding there, the population disperses, some birds staying to feed on nectar-poor flowers, others migrating elsewhere.

In temperate climates, however, where flowers die off in the fall, the entire population undertakes a true migration in a long-distance seasonal movement to a completely different home. Seen from an orbiting satellite with a telescope powerful enough to pick out tiny birds, the hummingbirds would show up as a swirling, intricate movement over the course of a year. For some species, the change would be slight. Anna's hummingbird is largely confined to the western seaboard of North America. It breeds in the Santa Monica Mountains in the winter, when gooseberries are in flower, and moves to mountain meadows in midsummer. There is a limited migration, as some birds move into Arizona and others wander as far north as Alaska and as far east as Florida.

Costa's hummingbird (Calypte costae) *migrates within the western United States between Colorado and California.*

Costa's hummingbird is another relatively sedentary species that nests in the Colorado Desert, where plants flower in winter and spring, then moves to the California coast. However, the two most familiar species of North American hummingbird, the ruby-throated and the rufous, are long-distance migrants that travel thousands of miles every year.

A much-discussed question among ornithologists is What is a migrant's "real" home? We tend to think of migrant birds as "living" in their summer homes, where they raise their families, and flying south only because conditions become too difficult in winter. Yet it can be argued that these birds are tropical or semitropical and fly north to take advantage of seasonal abundance for breeding. This would be true of the few hummingbird species that live outside the Tropics. The theory is supported by the example of the rufous hummingbird, which stays longer in its winter home than in its summer home (although it spends more time traveling than in either).

The impression that the winter home is not the regular residence is enhanced by the low status of visiting temperate hummingbirds among the southern residents. A study in the state of Catalina in Mexico showed that calliope and other hummingbird visitors have difficulty setting up feeding territories, and when they do, the territories are of poor quality, with few flowers. That these travelers can settle at all is due to an abundance of nectar flowers in winter. The resident hummingbirds breed in summer, when nectar is less abundant, so their population is kept relatively low, and there is a winter surplus that can be used by immigrants.

LONG-DISTANCE FLIGHTS

It is truly amazing that the hummingbirds feeding in North American yards in spring and summer will spend the winter thousands of miles to the south. And then one day in spring, the hummingbirds suddenly reappear, having come all the way back to exactly the same feeder or bed of flowers. The apparent impossibility of such small birds flying so far perhaps explains the popular myth that they migrate on the backs of Canada geese. The story probably began as a piece of folklore imported by settlers from Britain, where there was a similar story of the goldcrest (the smallest British bird) hitching rides on the backs of migrating owls.

Long-distance migration of birds has been studied intensively for more than half a century, and it is now common knowledge that birds

The ruby-throated hummingbird (Archilochus colubris) *is a long-distance migrant whose journey takes it as far north as southern Canada.*

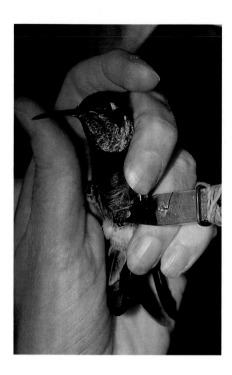

In order to better understand the migratory habits of hummingbirds, researchers band and track them.

have a remarkable ability to cover huge distances and to navigate with pinpoint accuracy. While it is not difficult to appreciate that large, powerful fliers such as geese, swans, cranes and even ducks and shorebirds can make intercontinental flights, it is not quite so easy to comprehend that this twice-yearly routine is also followed by small birds such as sparrows and finches. As for hummingbirds—such tiny scraps of life and tied to a lifestyle that requires frequent energy-rich meals— it seems inconceivable. But several species do indeed perform the wonder of long-distance migration. These are the species that nest the farthest north (the rufous, calliope and rubythroat) and the farthest south (the green-backed firecrown of southern South America).

Rufous hummingbirds nesting along the southern coast of Alaska regularly spend the winter in Mexico, some 2,000 miles away; a few stay in the very south of Texas and along the coast of the Gulf of Mexico. The calliope hummingbird travels much the same route and, at 0.1 ounce, is the smallest long-distance migrant in the bird world. Perhaps these feats are less surprising when we remember that hummingbirds regularly spend a large portion of the day in flight and must cover a fair number of miles to defend their territories and search for food, especially when feeding their young. A long journey in daily stages would not cause a great problem.

The details of the route taken by the rufous hummingbird are not well known, although this situation is changing as more hummingbirds are banded and recaptured on their travels. The first rufous hummingbirds arrive at their summer quarters while stragglers still remain in the Mexican wintering grounds. They set off from Mexico in February or March and travel up the West Coast, advancing as their favorite nectar flowers such as gooseberry, mountain larkspur and the tobacco plant (*Nicotiana*) come into bloom. They reach the breeding range in May, and the return journey starts in July. The southbound route takes the birds inland at heights of 12,000 feet or more in the Sierras and the Rockies. The birds stop at intervals and set up feeding territories that they maintain for a week or two while they refuel and build up their fat deposits.

When I watched the hummingbirds in Portal, I was struck by the aggression of the rufous hummingbird, which kept all comers from its feeder. I later learned that this is characteristic of the species. The rufous must compete with other hummingbird species it meets while migrating, and despite its small size, it succeeds because of its superior agility. It scores over other hummingbirds because it has relatively small wings. The drawback is that its flight is less efficient. Long-range migrants typically have long wings with a low wing loading (less weight relative to the wing area), so they stay airborne easily and fly economically. By breaking the journey into short legs between refueling stops, the rufous has traded the capability for economical long-range flight for agile flight that enables it to secure food.

Anna's hummingbird (Calypte anna) *has a limited migratory range.*

In the early days of birding, rufous hummingbirds would be spotted on rare occasions in winter along the Gulf Coast between Texas and Florida. In the past 20 years, these sightings have become much more frequent. The trend is not simply a result of people becoming more aware of wintering hummingbirds and recording them. Rufous hummingbirds are now regular winter visitors on the Gulf Coast, and their presence has extended inland some 250 miles into Georgia and Alabama. There has been a genuine change of behavior, perhaps because of an altered habitat. The Gulf Coast was once forest, which would not have provided winter food for hummingbirds, but flowering shrubs and herbs have now replaced the trees.

More recently, many more people have started growing flowering plants in their yards and installing hummingbird feeders. In every migrating bird species, there are rarities with errors in their genetically controlled navigation programs that fly to the wrong places. They provide excitement for birders but probably do not survive. When the Gulf habitat changed, the rare rufous hummingbird visitors flourished and passed on their "alternative" migration route and destination to their offspring, which in turn have been successful.

On the other side of the conti-

nent from the rufous, the ruby-throated hummingbird, in its long-distance flight to winter quarters in southern Mexico and Central America, is another example of stamina completely out of proportion to size. Rubythroats nesting in Canada and the central United States migrate in easy stages down through Texas and Mexico, but those in the eastern United States head due south until they get to the Gulf Coast. Details are sketchy, but it seems that many rubythroats continue around the Gulf Coast and down through Mexico; others pass down through Florida, then island-hop to Central America. But some make a prodigious leap across the Gulf of Mexico—about 500 miles. A few observations suggest that rubythroats take off well inland, where they have been seen gaining height in a steep climb until out of sight, but on their return, they have been spotted flying toward the coast low over the waves.

When researchers first discovered that rubythroats spend the winter in Central America, they believed that the birds must migrate over land. Apart from the apparent impossibility of such small birds surviving the long sea crossing over the Gulf of Mexico, the theory was that they could not perform such a flight because they did not carry enough fuel in the form of fat. It

Costa's hummingbird (Calypte costae) *migrates into the Colorado Desert when the desert vegetation comes into bloom.*

turned out that the theory was wrong.

The important point is that the distance a bird can fly does not depend on the absolute amount of fuel it carries but on the amount of fuel relative to its body weight. The smaller the bird, the higher the percentage of fuel it can carry. Small birds such as sparrows, finches and hummingbirds can carry 40 to 45 percent fat, which means they almost double their lean (fat-free) weight. (A large bird such as a swan or goose could not take off with this kind of burden.) The ability to carry a heavy load of fat puts the apparently supernormal stamina of small migrants into perspective. A ruby-throat with a lean weight of 0.1 ounce and carrying about 0.07 ounce of fat has fuel to spare for crossing the Gulf of Mexico by flying at an economical speed of 16 mph (more slowly than it moves from flower to flower when feeding).

The problem for rubythroats is adverse winds. Calculations show that they have fuel for a range of 1,400 miles in still air, which they are unlikely to encounter in the real world. Compared with large birds, hummingbirds are seriously hampered by headwinds. A headwind as slight as 10 mph will reduce the birds' range to 600 miles, and a 20 mph headwind will bring them to a

dead halt. On the other hand, a tailwind can sweep migrants along and carry them safely to their destination. In fact, there is evidence that many migrants wait for suitable weather before setting off on sea crossings.

TIMING

It is logical to assume that birds are driven away from their summer homes by worsening weather and food shortages in the fall. But it is not quite that simple: Male hummingbirds generally leave several weeks before females, and juveniles leave last. (The late departure of juveniles is evidence that migration is instinctive—young hummingbirds are not guided by their parents to their winter home.) In some places, male rufous hummingbirds vanish as early as the beginning of July, while some juveniles linger into November.

In the final analysis, it may be that birds migrate to escape the harshness of winter, but they are not forced into their journey by deteriorating conditions. That would be to court disaster—it is obviously suicidal to attempt a long journey without adequate preparations. Hummingbirds, like other migrants, must leave in good condition. Food supplies must be laid down in the form of fat to fuel the long flight, although it is

For northern hummingbirds, sap is an important food.

less necessary for birds that migrate in short stages and stop to replenish their reserves en route.

While a sudden change in the weather may precipitate the migrants' departure, this is only the final and not always necessary signal at the end of their preparations, which begin when the instinctive urge to migrate is triggered, probably by changes in day length. This is an important point, because many people who feed hummingbirds in their yards take down their feeders in the fall under the impression that an artificial food supply will prevent the birds from migrating and they will suffer when winter sets in. On the contrary, feeders may help birds build their reserves if nectar is scarce at this critical time, perhaps if rain or drought has ruined the flowers.

The fall migration is rather leisurely. Hummingbirds stop at frequent intervals and set up temporary feeding territories among late-blooming flowers. The ruby-throated hummingbird seems to follow specific flowers on its journey south, one of which is spotted jewelweed (*Impatiens capensis*). Interestingly, this species blooms several weeks later than the closely related *I. pallida*, a bee-pollinated flower.

By contrast, the spring migration is urgent; the birds hurry back to begin nesting. Some arrive on the breeding grounds even before the spring flowers have blossomed. They survive on insects and, these days, on sugar-water thoughtfully provided by householders. Sapsuckers also unwittingly provide an important source of nourishment.

Ruby-throated and rufous hummingbirds, the two most northerly species, are very dependent on the sap oozing from holes bored by sapsuckers. Coincidentally, the concentrations of sucrose and amino acids in sap are similar to those in nectar. It also seems to be no coincidence that the northern ranges of these two hummingbirds are similar to those of the yellow-bellied sapsucker.

STAY-AT-HOMES

Occasionally, a hummingbird stays behind in its summer home long after the others have gone and will perhaps remain for the entire winter season. A male broad-tailed hummingbird spent a winter in a front yard in Tucson, Arizona, instead of flying to Mexico. In addition to the lack of natural food, it had to cope with short daylight hours, only a third the length of those in summer, which reduced the time it had for feeding. With the aid of sugar-water from a feeder, small insects and nectar from the garden plant *Salvia greggii*, it stored enough food to avoid going torpid during the 14-hour nights. It even maintained enough strength to drive the larger resident Anna's hummingbirds from its territory.

Hummingbirds of several species now remain in the southern United States as late as December and, rarely, into early January before disappearing, perhaps taking a late migration rather than dying. But a few stay all winter. Rubythroats appear to be sensitive to cold and are rarely seen past October, except in very mild winters.

As with long migratory journeys,

cold weather is not such a great hazard for hummingbirds as we might think from their diminutive size and fragile appearance. In both cases, the key is to be well fed. Provided it has enough fat to maintain its body heat, a hummingbird can cope with freezing temperatures. The move to overwintering may be a response to global warming, but it is more likely to be the result of humans increasingly providing hummingbirds with winter food, either intentionally with feeders or unintentionally by growing beds of exotic flowering plants.

This theory might seem to contradict what was said earlier about feeders not causing hummingbirds to delay their migration. However, if a hummingbird has for some reason decided to prolong its stay through the winter, food provided by humans can help it survive bad spells. The truth is that this is not a subject about which much is known, but feeders do, at the very least, provide a means of locating late hummingbirds that would otherwise be overlooked. There are even stories of hummingbirds withstanding weather so cold that the sugar solution in the feeders froze. In a record three-day December freeze on the Gulf Coast, when temperatures plummeted to 11 degrees F, some hummingbirds survived without human help.

Finally, the role of insects in keeping hummingbirds alive in cold weather should not be overlooked. Insects sustain various small songbirds over winter, and they may be important for the hummingbirds that remain behind.

NESTING

Chapter Six

Compared with what we know of hummingbirds' nectar-sipping habits and their relationships with flowers, information about their reproductive life is limited, simply because it has not been as intensively studied. This is partly because of the difficulty of finding nests, especially those of species that nest in lush tropical vegetation or inaccessible sites. Indeed, the only nest of the blue-fronted lancebill (*Doryfera johannae*) ever found was in a mine shaft.

Nesting habits have also attracted less interest because they are remarkably uniform across hummingbird species. The study of hummingbird feeding behavior has been a fruitful field of research for ornithologists in providing insights into biological processes, but observations of nesting have yielded little more than simple descriptions of how hummingbirds time their breed-

Above left, a violet-headed hummingbird (Klais guimeti) *incubates her eggs. Right: Two Anna's hummingbird nestlings* (Calypte anna) *hardly fit into one nest.*

ing and rear their offspring. But as author Paul A. Johnsgard has pointed out, if it were not for a mating system in which males provide no parental support and can thus concentrate on courtship, hummingbirds "might well have been no more esthetically attractive than their drab relatives the swifts."

Compare hummingbirds with shorebirds, for example. Among approximately 190 species of shorebirds, breeding strategies range through all variations of parental care, from males simply mating and leaving females to rear the family, through monogamy, in which the parents share duties, to females courting males and leaving them to manage alone. Ornithologists have found intriguing lines of inquiry while studying the benefits and drawbacks of each type of parenthood.

Hummingbirds, however, show little variety in their reproductive behavior. Generally, the male stays with the female only long enough for fertilization then leaves her to incubate the eggs and rear the nestlings on her own. Indeed, in no other large bird family has there been such a well-defined and universal separation of males from parental responsibilities.

A few reports exist of male hummingbirds incubating eggs and, less frequently, feeding young, but they are tantalizingly rare and incomplete. On the basis of two brief accounts, the sparkling violetear has been cited as the exception to the rule. One report is of a male taking over incubation when the female was killed; the other is of two adults sharing incubation at the nest, one

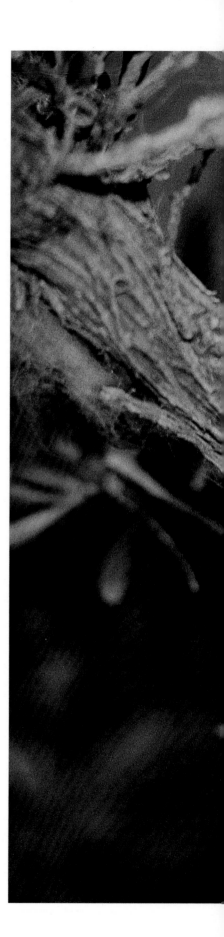

A white-eared hummingbird (Hylocharis leucotis) *works a tangle of fibers into place as she builds her nest.*

of which the observer believed to be a male. The plumage of the two sexes is similar in this species, so it is very slender evidence; it is also known that in other bird families, two females sometimes share a nest. Subsequent studies have failed to reveal any unusual male habits.

Among North American species, male ruby-throated and rufous hummingbirds have been observed sitting on the nest, and one male Anna's has been seen feeding young. The title "Anna's Hummingbird in Adult Male Plumage Feeds Nestling" in the journal *Condor* suggests that the birds in question may not actually be males. Female sunangels (*Heliangelus*) have varying amounts of iridescent feathers on their gorgets, and some glitter as much as the males do. If similar variation occurs in other species, it is possible that the "male" hummingbirds were physiological anomalies (females in male plumage), not behavioral ones (males performing a traditional female function).

If these birds were actually males, it raises the question of what made them break the rules of hummingbird behavior. For four years, Alexander Skutch watched a male bronzy hermit (*Glaucis aenea*) at his home that kept a female company throughout her nesting. Not only did the male inspect the nest, but in the early stages, he sat in it and went through the motions of building. Perhaps the only conclusion to draw from these examples is that hummingbirds, and other birds, are not perfectly programmed automatons; rare individuals can exhibit aberrant behavior.

BREEDING SEASONS

Hummingbird breeding seasons coincide with a plentiful supply of food, which is the flowering season for their main nectar flowers. Temperate species nest in the northern or southern spring and summer when favorite nectar flowers are in full bloom. The nesting of the ruby-throated hummingbird may be correlated with the flowering of wild columbine (*Aquilegia canadensis*), but there are some interesting variations. Anna's hummingbird is the earliest nester of any North American bird. In the chaparral of California, it begins in December or even November, the wettest part of the year, when fuchsia-flowered gooseberry (*Ribes speciosum*) blooms.

In the Tropics and Subtropics, eggs may be laid at any time of year, but there are sometimes two peaks of laying. The rainy season sees the greatest profusion of flowers, but hummingbirds nest mainly at the beginning and the end, when there is not enough rain to swamp their tiny, exposed nests. Hermit hummingbirds build their nests under leaves

Early-flowering gooseberries provide the food that enables Anna's hummingbird (Calypte anna) *to nest in winter.*

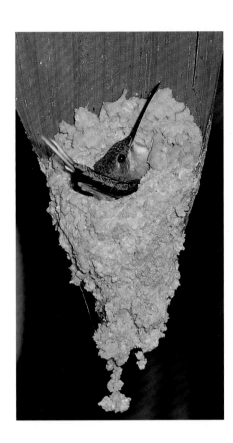

The straight-billed hermit (Phaethornis bourcieri) *builds its pendulous nest on the tip of a rainforest leaf.*

for protection, so nesting can continue throughout the rainy season. The weight of the long-tailed hermit and its nest is sufficient to bend the supporting palm frond so that it provides better protection. Some hermits in lowland tropical forests build loosely woven nests that allow water to drain away quickly.

Local conditions give rise to considerable variation in breeding season. The white-eared hummingbird, for instance, nests in Guatemala when dry-season flowers are beginning to blossom. The green violetear nests in central Mexico at the end of July and raises one family, but those nesting in Costa Rica raise two families between October and March. The rufous-tailed hummingbird nests throughout the year except while molting. The rose-throated hummingbird (*Selasphorus flammula*) lives in the mountains of Costa Rica, where it nests at altitudes as high as 9,600 feet. The cold nights and thin air (which makes flying more difficult) increase the need for a rich daytime source of nectar, and the bird synchronizes its breeding with the short blooming season of local *Salvia* and *Tropaeolum* species, whose flowers have unusually high concentrations of nectar.

Among migratory species, the males arrive in the summer breeding grounds before the females—by as much as several weeks for rufous hummingbirds. This is a common pattern among migrating birds; the male's priority is to set up a territory before all the vacant lots are taken. The females follow at a leisurely pace and arrive when there is plenty

of food to provide raw materials for manufacturing eggs. Many begin nest-building as soon as they arrive.

NESTING

Before expending any energy and enthusiasm on courtship, the female hummingbird quietly builds her nest. Although the nest is often no more than a simple cup, it is still a remarkable construction that takes several days to complete, sometimes as long as two weeks. Perhaps because hummingbirds invest so much time and energy in nest-building, they often refurbish old nests, sometimes ones they did not build themselves. The main structure of the nest is made of moss, grass, buds, rootlets, cotton from seeds, down from woolly leaves and animal hair. Sometimes the lining is downy; the exterior may be camouflaged with bits of lichen, moss and bark. The whole edifice is anchored and bound together with spiders' silk, nectar or saliva. The Andean hillstar uses nectar to glue its nest to a rock overhang or to the ceiling of a cave; some species prefer to build nests on branches overhanging water. The main criterion for a nest site, however, is a rich source of nectar nearby.

The most basic nests are simple cups saddled on top of a branch or across a fork, but a few species, including the sylphs (*Aglaiocercus*), make domed nests, and high-altitude species make bulkier nests that provide better insulation. The hermits make elaborate nests that hang beneath large leaves and palm fronds. The nest of the long-tailed hermit is built as a bracket against the surface

A black-chinned hummingbird
(Archilochus alexandri) *nestling breaks*
its way into the world; the second egg
will hatch shortly afterward.

of a frond, and the sooty-capped hermit (*Phaethornis augusti*) builds a remarkable construction in which the cup hangs by a single stout rope of spiders' web attached to one side of the rim. Directly below the nest is a second cable weighted with pebbles and lumps of clay that counterbalances the lopsided structure.

Nests are protected from rain and midday sun by overhanging rocks or vegetation. But protection is also needed at night, when radiant heat can be lost to a clear sky. This is particularly important for hummingbirds nesting in cool places; they need additional shelter to augment the insulation provided by the nest. An Anna's hummingbird in the Santa Catalina Mountains in California maintained the temperature of her eggs at a temperature 15 Fahrenheit degrees higher than that of the air. It helped that the nest was in a steep-sided canyon whose walls radiated the warmth they had stored during the day. Shelter is even more important for hummingbirds nesting at high altitudes. The Chimborazo hillstar, which lives just below the snow line on the volcano Cotopaxi in Ecuador, builds its nest in shallow caves on the sides of ravines. The Andean hillstar and the calliope hummingbird, which nest at high altitudes, choose east-facing nest sites

The young of the rufous hummingbird (Selasphorus rufus) *receive a mixture of nectar and insects from their mother.*

that catch the rays of the rising sun.

The female lays no more than two pure white eggs, and only rarely one. An extra egg or two appearing in a nest most likely belongs to a second female. The size of bird eggs is roughly proportional to body size, so it is not surprising that hummingbirds lay the smallest eggs of any bird. An egg of the bee hummingbird, the smallest species, in the collection of the U.S. National Museum in Washington, D.C., and apparently the only known specimen, is about ½ inch long and ⅓ inch in diameter and weighs 0.018 ounce. The giant hummingbird's egg is the largest, averaging ⅘ inch by ½ inch.

Some species begin incubation when they lay the first egg, others when they lay the second. Costa's, ruby-throated and rufous hummingbirds start with the first egg; the eggs hatch one to three days apart, depending on the interval between laying. The eggs of calliope hummingbirds, however, hatch at the same time, because incubation begins only after the second egg is laid. Incubation times are relatively long, usually 15 to 17 days and sometimes as long as three weeks in mountain species. This may be because the females have to leave the nest frequently to feed; eggs develop more slowly when they cool. The long-tailed hermit incubates for periods of 45 minutes, with intervals of 25 minutes for feeding and collecting extra nest material and spiders' silk for renewing the bindings.

Objects of small volume cool (and heat) quickly, so hummingbird eggs are more vulnerable to cooling than those of most birds. Females incubating eggs cannot go into torpor because they must maintain a high body temperature to keep the eggs warm. They also feed later in the evening in order to retire well stocked with food to carry them through the night. Although the female continues to feed through the day, her foraging bouts are widely spaced, and she economizes by reducing her body temperature by about 15 Fahrenheit degrees, achieving a 50 percent energy saving.

The danger of cooling is more critical for species at higher altitudes and latitudes. Hummingbirds in the high Andes build deeper, thicker nests than those at lower altitudes. Rufous hummingbirds nesting in Canada and Alaska vary their nest sites to take advantage of changing weather through the nesting season. In spring, they prefer to nest at low levels, where they are protected from severe weather, but later nests are built in broad-leaved trees whose sheltering foliage prevents overheating.

When the nestlings hatch, naked and helpless, their mother broods

A broad-tailed hummingbird (Selasphorus platycercus) *nest is decorated with lichens, perhaps for camouflage.*

Once it leaves the nest, a black-chinned hummingbird (Archilochus alexandri) *will continue to be fed by its mother for a few more days.*

them to keep them warm until they are able to maintain and regulate their body temperatures, which usually takes a week or more. When the nest is in an exposed position, the female may have to sit over them to shade them from excessive heat. Even though they are as susceptible to rapid cooling as the eggs, hummingbird nestlings are unusual among birds in that they do not develop a layer of down before growing their main plumage. Nestlings of Anna's hummingbird have been described as transforming from "loathsome black worms into tiny porcupines" when their feathers first sprout.

Shortly after the nestlings hatch, the mother begins delivering frequent meals to them. At first, she stimulates them to gape by prodding them on the head with the tip of her bill, then she pumps nectar from her crop or drops insects from the tip of her bill into the waiting mouths.

By the time they are about a week old, nestlings respond to the breeze from her wing beats as she hovers near them. A well-fed nestling is identifiable by a large bulge in its throat. Protein-rich insect food is important for growing nestlings, and Anna's hummingbird stops defending flowers and spends more time hunting insects when her nestlings hatch. She also feeds them more insects in the afternoon, perhaps because insects are relatively inactive and harder to find in the morning or because protein is a long-lasting food reserve that will maintain the nestlings overnight.

The nestlings gradually outgrow the nest. Most first leave after 23 to 26 days, depending on species and the abundance of food; Andean hummingbirds stay in the nest for 30 to 40 days. The young hummingbirds make their first flights before their wing and tail feathers are fully grown, and they continue to receive food from their mother for three weeks or more. When the fledglings first leave the nest, they stay in one place and make begging calls to attract her attention. Later they follow her, and she calls them to be fed. During this time, they learn where to look for nectar by watching their mother and by trial and error as they investigate and learn to reject unlikely sources such as leaf surfaces.

LONGEVITY AND MORTALITY

As with many small birds, losses over the nesting period are high. Many clutches of eggs are lost because of bad weather. In drought years, nectar can run short, and birds die of starvation; a cold spell can kill adults on the nest. About a third of the losses in the Tropics are through predation, a figure that rises to two-thirds in North America. Nests are destroyed by a variety of nest robbers such as jays, tanagers, squirrels and snakes. Fire ants also invade nests and consume nestlings. In one case, a brown-headed cowbird egg replaced the original clutch in a ruby-throated hummingbird's nest. Nest destruction occurs despite valiant attempts at defense by the parent. Some hummingbirds merely scold the robbers, but others actively attack and buffet them with their wings, fearlessly mobbing owls, squirrels, eagles and humans.

Little is known about predation

on adult hummingbirds, except that they fall victim to a number of animals. No predators are known to specialize on them, however, with the possible exception of the eyelash palm pit viper, which lies in wait on flowers for visiting nectar seekers. No doubt hummingbirds are often saved by their speed and maneuverability; attacks are likely to be successful only when the birds are taken by surprise. Small hawks and owls occasionally catch them, and there are unusual accounts of capture by frogs, praying mantises, dragonflies, spiders and even a bass that leaped out of a pond and swallowed a hummingbird. Accidents can befall such small animals as well. They can be trapped in spiders' webs and bristly seedheads, and human artifacts such as cars, windows and mesh screens are death traps.

Despite such dangers, the life expectancy of a hummingbird is good for a small bird. There have been few studies of survival and longevity, even for common species, but most experts think their average lifespan is three or four years. The record holder is a female broad-tailed hummingbird, banded as an adult in Colorado in 1976 and recaptured in the same location in 1987, which would mean she was at least 12 years old.

Hummingbirds breed in their second year, and some produce two or three clutches a year, so the potential for raising enough offspring to maintain the population is good. But as the next chapter will show, the catastrophic effects of human activity may mean that it is not enough.

No known predators specialize in adult hummingbirds, with the possible exception of the eyelash palm pit viper, here poised to attack an unsuspecting rufous hummingbird (Selasphorus rufus).

HUMMINGBIRDS AND HUMANS

Chapter Seven

Hummingbirds have been objects of fascination since the time humans first arrived in the Americas by crossing the Bering Strait. Although larger, more edible and more dangerous animals were plentiful, the hummingbird's size, gorgeous colors and clear differences from other birds made it a favored symbol in myth and religion. Many hummingbird stories disappeared centuries ago, but echoes of their original impact still exist in pre-Columbian artifacts and artwork and in tales that survived the arrival of Europeans. Meanwhile, new relationships with hummingbirds are being established through the popular practice of attracting them to the home with feeders. Like many forms of wildlife, however, hummingbirds are under pressure as their habitats are transformed and ruined.

Above left, a vase created by the Nazca, whose art and architecture were inspired by hummingbirds such as the ruby topaz (Chrysolampis mosquitus), *right.*

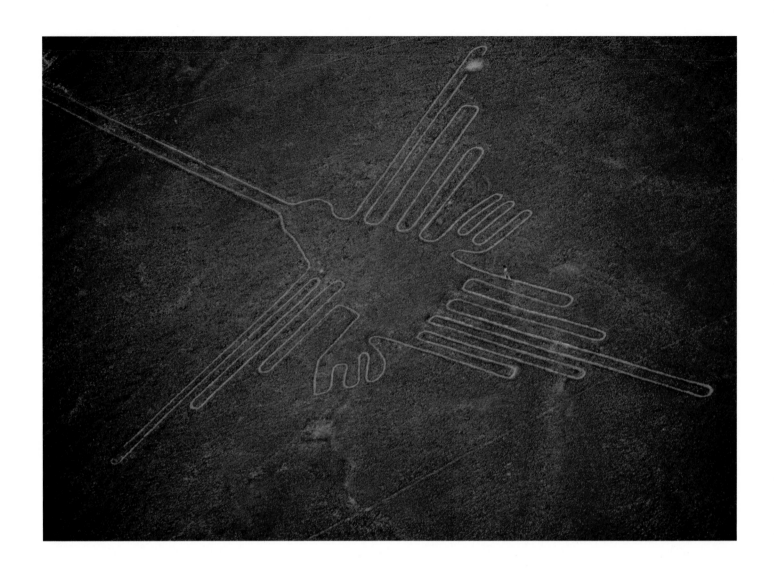

One of some 30 controversial giant
outlines of animals and other life forms
known to date from the time of the
Nazca (200 B.C.–600 A.D.), this huge
hummingbird drawing stretches 300 feet
across the Peruvian plateau.

HUMMINGBIRDS IN HISTORY

Records of hummingbirds in past civilizations exist in the stone carvings of the Aztec and in the huge 300-foot-long hummingbird outline carved into the surface of a plateau by the Nazca, who flourished in Peru about 1,500 years ago. Thousands of miles north, hummingbirds are depicted among the hundreds of ancient rock carvings at Three Rivers in New Mexico. We can only speculate about the significance of such artifacts. More is known about Aztec and Mayan hummingbird myths, because the Spanish and other Europeans recorded them before destroying those civilizations.

A drawing from 1579 shows the Aztec war god Huitzilopochtli. Originally a human warrior, he was killed by an arrow then immediately transformed into a spirit hummingbird and given his name, which means Hummingbird of the South. Eventually, all fallen warriors became hummingbirds and were charged with defending the nation under Huitzilopochtli. They fed on nectar and practiced their fighting skills by attacking each other whenever they met. They were required to be particularly vigilant at dawn and dusk to stave off the powers of darkness. (It is easy to see how the myth could arise from watching hummingbirds.) Eventually, the Aztec king Huitzilihuite (Hummingbird Feather) made Huitzilopochtli the supreme sun god, while the goddess of beauty and birth was Xochiquetzal (Flower Bird).

According to Mayan myth, the creator made all the other birds before the hummingbird. A few scraps were left over, and from these, he fashioned a pair of tiny birds—male and female—with long bills that were perfect for getting nectar from flowers. But the creator had to give them the power to hover so that they could feed at the flowers. The buzzing of their hovering flight earned the male the name Dzunuume (Hummingbird). When the two hummingbirds came to be married, the other birds took pity on them because of their drab feathers. The quetzal gave them the shimmering green of its tail, the swallow gave them white feathers from its breast, and the house finch gave them red feathers for their throats. The male hummingbird got most of these. Then the sun blessed the pair with its rays to make their feathers shine as long as they faced it. So affected was Dzunuume by all this attention that he virtually lost his voice and was left with only a feeble song.

For many peoples, the hummingbird was the bringer of rain. The Aztec, the Toltec, the Hopi, the Zuni and others have myths of hummingbirds bringing rain after a drought. The Pima of the southwestern United States told of a terrible drought caused by the rain and wind spirits becoming angry and hiding in their sacred cave. Neither prayers nor dances brought relief to the withered crops, and both humans and animals were sent to find the spirits, but without success. Eventually, despite everyone's scorn, the tiny hummingbird volunteered. Taking a magic thong made from the hair of the Pima chief's daughter, he flew to the cave and bound the spirits as they slept. When they awoke, he told

A Nazca wooden carving in the shape of a hummingbird.

them they would be released only if they promised to return with him to the land of the Pimas, which is why wind, rain and hummingbirds arrive together in this part of the country. The Hopi and Zuni peoples still imitate hummingbirds in their rain dances.

Some of the old symbolism of the hummingbird has survived into modern life. Dead, dried hummingbirds are worn as good-luck charms around the neck or displayed in cars. In Mexico, powdered hummingbird is a love potion administered to the object of passion to ensure constancy.

In 1492, when Christopher Columbus first arrived in the Americas, his initial contacts were with the Taino people of the Caribbean and Florida. They believed that hummingbirds were once small flies the sun god had changed into little birds; they symbolized life and rebirth. The Taino also called their warriors *colibri* warriors, or hummingbird warriors, because although hummingbirds are usually peaceful, they fiercely protect their homeland. *Colibri* has passed into Spanish, French, Portuguese and Italian as the word for hummingbird, and the Germans have modified the spelling to *Kolibri*. (The English came to know hummingbirds when they met the rubythroat in New England and named it for the sound of its wings.)

Columbus and other visitors reported hummingbirds among the natural marvels of the New World, and soon hummingbird skins, their wonderful colors undimmed, were being sent back to Europe. One reached the Pope in 1516, but it was nothing compared with the ceremonial headdress of the Aztec king Montezuma, which featured several hundred skins. Montezuma gave the headdress to his conqueror Hernán Cortés, who then presented it to Emperor Charles V. It is now in the Museum für Völkerkunde in Vienna.

By the late 17th century, John Bannister of Virginia was shipping kiln-dried hummingbirds to England for sale at £8 each. But the heyday of the hummingbird trade came in the 19th century. Tens of thousands of hummingbirds passed through the hands of London dealers every month; one dealer imported 400,000 in a year. The hummingbirds' fate was to be curios in glass cases, or their feathers were incorporated into bizarre drawing-room ornaments such as artificial flowers and pictures or used as decorations for ladies' hats. Recommending collecting stuffed hummingbirds as a hobby suitable for ladies, Adolphe Boucard wrote, "It is as beautiful and much more varied than a collection of precious stones and costs much less."

Eventually, revulsion for this use of birds caused the trade in skins and feathers to dwindle and finally die

According to legend, fallen Aztec warriors were transformed into hummingbirds. Left: the white-vented violetear (Colibri serrirostris).

Hummingbird feathers were once a popular fashion accessory on hats and dresses.

out. Most hummingbird populations recovered when the persecution stopped, although it is likely that a few species became extinct through the trade—some have not been seen, alive or dead, for a century or more.

While dead hummingbirds were being transported across the Atlantic, there were also attempts to bring back live birds for exhibition and for sale to aviculturists. John Gould managed to bring one live hummingbird back to London, but it died two days after its arrival. Improved diet and faster transport brought more success in later years, but the death rate of captives remained high. Even when the importance of protein in hummingbirds' diet was understood, it was difficult to keep them in captivity because of the various species' differing requirements. Finding the correct formula for artificial diets has remained a problem for aviculturists.

CONSERVATION

Attitudes toward wildlife have changed over the past century. When John Gould came over from England in 1857, he was delighted to see his first live hummingbird. He later wrote, "With what delight did I examine its tiny body and feast my eyes on its glittering plumage." This suggests that he shot it to get a closer look, not unusual behavior at the time. In fact, it is only relatively recently that ornithologists have not shot birds to confirm their identity.

Attitudes have also changed toward trade in birds. Compared with the situation 100 to 150 years ago, when thousands of dead humming-birds were sent to Europe, it is now difficult to export live hummingbirds from their native countries. Since 1987, all hummingbird species are in Appendix II of the Convention on International Trade in Endangered Species (CITES), classified as being "of conservation concern." This means that export is not allowed without a permit. In practice, permits are rarely granted because almost every country inhabited by wild hummingbirds has its own legislation forbidding their export.

Like other endangered and threatened species, the greatest danger now facing hummingbirds is habitat destruction. Environmental degradation and devastation are proceeding apace through much of the Americas. However, it is not easy to calculate the population of a species in a remote location let alone monitor whether its level has fallen. The Honduras emerald, for example, was known from only 11 specimens before 1950, but now that ornithologists have penetrated its remote home in northern Honduras, it has been found to be common in several valleys. BirdLife International, a global partnership of conservation organizations based in Cambridge, England, recognizes 27 hummingbird species as "globally threatened" and another 22 as "near threatened." Of the 27 globally threatened species, 8 are classed as "critically endangered," 7 as "endangered" and 12 as "vulnerable." One species, Brace's emerald (*Chlorostilbon bracei*), is definitely extinct. It is known only from a single specimen collected in 1877 on New Providence

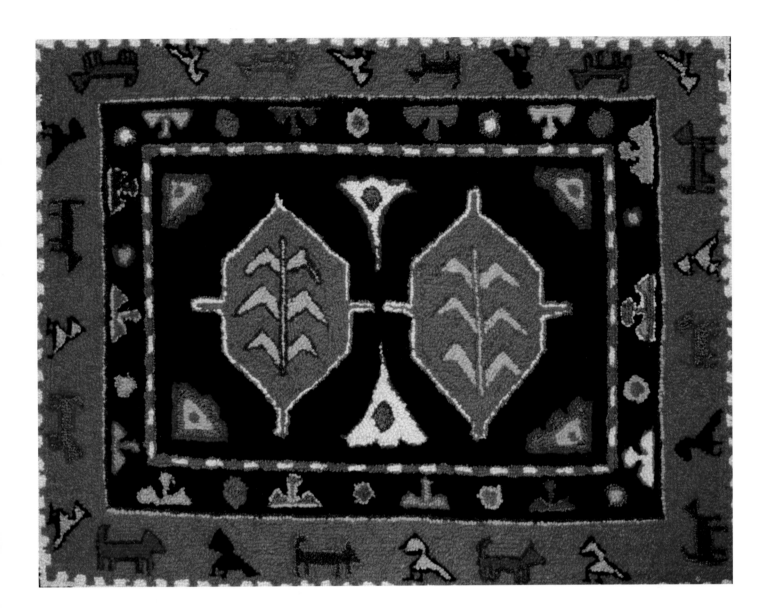

Incorporated into a modern hooked rug, an Aztec design reflects that ancient civilization's choice of the hummingbird as one of its cultural symbols.

Island in the Bahamas. Another, the coppery thorntail (*Popelairia letitiae*), is known only from three specimens, and as it has not been seen for 100 years, it is probably extinct as well.

The dangers to threatened hummingbird species are consistent with the conditions that threaten so many birds and other animals in Central and South America, namely development in the form of logging, mining, clearance for agriculture and road- and house-building, some of which take place even in national parks. Some hummingbirds can cope with considerable change. Those that live at forest edges or feed on flowers in newly opened country may find new opportunities. Others survive in banana and coffee plantations or have adapted to urban life.

The species most at risk are those with specialized habitats and diets or restricted ranges. They face the danger that their entire habitat may be destroyed or broken into fragments too small to support viable populations. There are, for instance, probably fewer than 250 (and perhaps as few as 50) individuals of the hook-billed hermit (*Glaucis dohrnii*) remaining in 40 square miles of scattered fragments in the seriously threatened lowland Atlantic forests of eastern Brazil. In 1939, this species would have been

Although hummingbirds no longer visit the ruins of Tzintzuntzan in Mexico, it is said that its name means "abundant in hummingbirds."

abundant over a range of 13,500 square miles.

Other species, however, inhabit ranges that were never extensive. This is particularly true for hummingbirds living on islands. The Juan Fernandez firecrown (*Sephanoides fernandensis*) has disappeared from some of the islands in the Chilean archipelago, and only a few hundred remain. The forest is disappearing, and the firecrown is also threatened by cats, rats and other introduced mammals as well as by competition with the mainland green-backed firecrown. "Habitat islands" such as patches of forest surrounded by grass are as dangerously limited as those surrounded by water. The home of the black-breasted puffleg (*Eriocnemis nigrivestis*) is on mountain peaks above 9,000 feet in northwest Ecuador, but it has disappeared from all except the volcano Pinchincha, where fewer than 50 now survive.

A less obvious danger faces migratory hummingbirds. Some perform regular movements around their range, visiting a mosaic of habitats to take advantage of the various blooming times of different assemblies of flowers. If one of these crops fails and there is no alternative link in the annual food cycle, the species is doomed as surely as if its entire habitat were destroyed. It is a challenge for conservationists to discover the complete requirements of a species and implement a rescue plan that entails preserving a section of every one of its habitats.

The hummingbirds of North America are faring better than their southern relatives, although the danger to the long-distance travelers—rufous, calliope and rubythroat—is that their winter quarters and staging posts are being destroyed. In their North American summer home, there is some loss of habitat due to development, but the practices of planting hummingbird flowers and providing sugar-water feeders in yards help offset its impact. Fires and clearance for agriculture are local problems, but probably only Costa's hummingbird is of conservation concern as its habitat on the desert fringes of the southwest comes under increasing pressure.

Like the majority of the world's bird species, hummingbirds face an uncertain future as the human population continues to make further inroads into the wilderness. Species that can adapt to the changes have the best chance of survival. Hummingbirds, despite the specialized feeding habits that make them unique objects of wonder and interest, appear to have a brighter future than many other birds.

Costa Rica's green-crowned brilliant hummingbird (Heliodoxa jacula) *is an important pollinator of flowering plants in wet highland regions.*

Further Reading

Bent, A.C. *Life Histories of North American Cuckoos, Goatsuckers, Hummingbirds and Their Allies.* Smithsonian Institution, 1940.

Del Hoyo, J., Elliot, A. and Sargatal, J. (eds). *Handbook of Birds of the World.* 6 volumes, Lynx Edicions, 1992.

Gould, J.A. *Monograph of the Trochilidae, or Family of Humming-birds.* Taylor and Francis, 1861. Reprinted as *Hummingbirds.* Wellfleet Press, 1990.

Grant, K.A., and Grant, V. *Hummingbirds and Their Flowers.* Columbia University Press, 1968.

Greenewalt, C.H. *Hummingbirds.* Doubleday and AMNH, 1960. Dover, 1990.

Holmgren, V.C. *The Way of the Hummingbird in Legend, History and Today's Gardens.* Capra Press, 1986.

Johnsgard, P.A. *The Hummingbirds of North America.* Smithsonian Institution, 1997.

Scheithauer, W. *Hummingbirds: Flying Jewels.* Translated from Kolibris by G. Vevers. Arthur Barker, 1967.

Schneck, M.H. *Creating a Hummingbird Garden.* Fireside, 1994.

Skutch, A.F. *The Life of the Hummingbird.* Vineyard Books, 1973.

Stokes, D. and L. *The Hummingbird Book.* Little, Brown & Co., 1989.

Photo Credits

67: © Kenneth W. Fink/
   Bruce Coleman Inc.
68: © Michael and Patricia Fogden
70: © Michael and Patricia Fogden
72: © Michael and Patricia Fogden
75: © Joe McDonald/DRK Photo
76: © Michael and Patricia Fogden
78: © Richard K. LaVal/
   Animals Animals
79 top: © Arthur Gloor/
   Animals Animals
79 bottom: © David Welling/
   Animals Animals
80: © Michael and Patricia Fogden

Chapter Four/Daily and Social Lives
82: © Tim Fitzharris
83: © Luis Mazariegos
84: © Tim Fitzharris
87: © Michael and Patricia Fogden
88: © Sid and Shirley Rucker/
   DRK Photo
91: © Michael and Patricia Fogden
92: © Luis Mazariegos
94: © Luis Mazariegos
97: © Michael and Patricia Fogden
98: © Tim Fitzharris
100: © Stephen Dalton/
   Animals Animals/Earth Scenes

Chapter Five/Tiny Travelers
102: © Maresa Pryor/
   Animals Animals
103: © Barbara Gerlach/DRK Photo
104: © Tim Fitzharris
107: © Stephen J. Krasemann/
   DRK Photo
108: © John Cancalosi/DRK Photo
109: © Kenneth W. Fink/
   Bruce Coleman Inc.
110: © Lisa Husar/DRK Photo
113: © Sid and Shirley Rucker/
   DRK Photo
114: © Jack Wilburn/Animals Animals

Chapter Six/Nesting
116: © Michael and Patricia Fogden
117: © Mark A. Chappell/
   Animals Animals
119: © Phil Degginger/
   Animals Animals
120: © John Cancalosi/DRK Photo
122: © Michael and Patricia Fogden
123: © Sid and Shirley Rucker/
   DRK Photo
124: © Tom and Pat Leeson/
   DRK Photo
127: © Jeff Foott/DRK Photo
128: © Sid and Shirley Rucker/
   DRK Photo
130: © Michael and Patricia Fogden/
   DRK Photo
131: © Michael and Patricia Fogden/
   DRK Photo

Chapter Seven/Hummingbirds and Humans
132: © Gianni Dagli Orti/CORBIS
133: © Luis Mazariegos
134: © Yann Arthus-Bertrand/CORBIS
135: © Gianni Dagli Orti/CORBIS
136: © Luis Mazariegos
138: © Hulton-Deutsch Collection/
   CORBIS
139: © Jacqui Hurst/CORBIS
140: © Danny Lehman/CORBIS
143: © Michael and Patricia Fogden

Back pages
145: © Luis Mazariegos
147: © Luis Mazariegos
151: © Stephen J. Krasemann/
   DRK Photo

Index